A handbook of Fourier theorems

A HANDBOOK OF

Fourier theorems

D. C. CHAMPENEY

School of Mathematics and Physics, University of East Anglia

The right of the University of Cambridge to print and sell all manner of books was granted by Henry VIII in 1534. The University has printed and published continuously since 1584.

CAMBRIDGE UNIVERSITY PRESS

Cambridge

New York Port Chester Melbourne Sydney

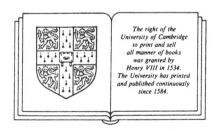

Published by the Press Syndicate of the University of Cambridge
The Pitt Building, Trumpington Street, Cambridge CB2 1RP
40 West 20th Street, New York, NY 10011, USA
10 Stamford Road, Oakleigh, Melbourne 3166, Australia

First published 1987
First paperback edition 1989
Reprinted 1990

British Library cataloguing in publication data

Champeney, D. C.
A handbook of Fourier theorems.

1. Fourier analysis
I. Title
515′.2433 QA403.5

Library of Congress cataloguing in publication data

Champeney, D. C.
A handbook of Fourier theorems.

Bibliography
Includes index.
1. Fourier analysis. I. Title.
QA403.5.C47 1987 515′.2433 86-32694

ISBN 0 521 26503 7 hard covers
ISBN 0 521 36688 7 paperback

Transferred to digital printing 2000

MP

To Julie, John and Anna

CONTENTS

PREFACE

This handbook is intended to assist those scientists, engineers and applied mathematicians who are already familiar with Fourier theory and its applications in a non-rigorous way, but who wish to find out the exact mathematical conditions under which particular results can be used. A reader is assumed whose mathematical grounding in other respects goes no further than the traditional first year university course in mathematics taken by physical scientists or engineers. Advances in mathematical sophistication have led to a growing divide between those books intended for mathematics specialists and those intended for others, and this handbook represents a conscious effort to bridge this gap.

The core of the book consists of rigorous statements of the most important theorems in Fourier theory, together with explanatory comments and examples, and this occupies chapters 6–16. This is preceded, in chapters 1–5, by an introduction to the terminology and the necessary ideas in mathematical analysis including, for instance, the interpretation of Lebesgue integrals. Proofs of theorems are not provided, and the first part is not intended as a complete grounding in mathematical analysis; however, it is intended that the book should be self contained and that it should provide a background which will assist the interested reader to follow proofs in standard mathematical texts.

In the chapters on Fourier theorems, those classical theorems dealing with locally integrable functions are covered first in a way that leads naturally to the later sections covering generalized functions. The more modern extensions in the direction of abstract harmonic analysis are not covered.

PREFACE

[illegible]

1

Introduction

Three of the most important theorems in the theory of Fourier analysis are the inversion theorem, the convolution theorem, and the differentiation theorem. These may be stated in a rough and ready way as follows. Given functions $f(x)$ and $g(x)$, and defining their Fourier transforms $F(y)$ and $G(y)$ by

$$F(y)=\int_{-\infty}^{\infty} f(x)\exp(-2\pi \mathrm{i}xy)\mathrm{d}x$$

and

$$G(y)=\int_{-\infty}^{\infty} g(x)\exp(-2\pi \mathrm{i}xy)\mathrm{d}x,$$

then it follows that:

inversion theorem

$$f(x)=\int_{-\infty}^{\infty} F(y)\exp(2\pi \mathrm{i}xy)\mathrm{d}y \tag{1.1}$$

convolution theorem

$$\int_{-\infty}^{\infty} f(u)g(x-u)\mathrm{d}u=\int_{-\infty}^{\infty} F(y)G(y)\exp(2\pi \mathrm{i}xy)\mathrm{d}y \tag{1.2}$$

differentiation theorem

$$\text{if } g(x)=\frac{\mathrm{d}f}{\mathrm{d}x} \text{ then } G(y)=2\pi \mathrm{i}yF(y).$$

The above statement of the inversion theorem is ambiguous in several ways including the following: (i) it does not state to what class of function $f(x)$ the theorem applies, (ii) it does not state whether the integrals are to be interpreted as Riemann, Lebesgue, generalized, or some other form of integral, and (iii) in cases when (1.1) applies to some, but not all, values of x

it is not clear at what values of x the equation will apply or in what sense convergence of the integral to the function $f(x)$ occurs. The statements of the other two theorems are also ambiguous in several respects.

The aim of this book is to present rigorous statements of these theorems, and of several others, together with sufficient background material to make the statements intelligible. We assume a reader who is already fairly familiar with the applications of Fourier theory but whose pure mathematics does not reach beyond that found in a typical undergraduate degree course in a physical science. A knowledge of Lebesgue integration is not assumed at the outset.

The life and work of Jean Baptiste Joseph Fourier (1768–1830) is described in books by Grattan-Guinness (1972) and by Herivel (1975), whilst Fourier's *Analytical Theory of Heat* is available in English translation (Fourier, 1955), and it is clear from these sources that Fourier himself did not give precise sets of conditions under which the inversion theorem would hold. His achievement, and deserved fame, lay in discovering and demonstrating how the theorem could be usefully applied in physical problems, and it fell to subsequent mathematicians to add rigour to Fourier's ideas. Thus it is that 'the Fourier theorem' consists not in one single theorem, but in several theorems all on a common theme.

Much depends upon how one interprets the integral symbol in these theorems. One needs first to decide whether Riemann or Lebesgue integration is intended. If, as is almost always the case in modern treatments, Lebesgue integration is used one then needs to know whether the symbol $\int_{-\infty}^{+\infty} f(x)\mathrm{d}x$ is to be interpreted as a direct Lebesgue integral or as some limit such as $\lim_{\lambda\to\infty}\int_{-\lambda}^{+\lambda} f(x)\mathrm{d}x$ or $\lim_{a\to 0}\int_{-\infty}^{+\infty} \mathrm{e}^{-a|x|}f(x)\mathrm{d}x$, or in some other way such as a limit in the mean, described in chapter 4. Each interpretation brings with it its own theorem. Indeed when one recognizes that the meaning of the integral symbol can be subject to doubt one realizes that the very definition of what is meant by a Fourier transformation becomes uncertain. We start in chapters 6 and 7 with the simplest definition and follow this with others of increasing sophistication in later chapters.

Because of the central role of integration, we commence, in chapter 2, by describing Lebesgue integration, following this by some useful associated theorems in chapter 3. Chapter 4 then deals with the convergence of sequences of functions. Sequences are important because of the following question: if the functions $f_1(x)$, $f_2(x)$, $f_3(x)$, ... have Fourier transforms $F_1(y)$, $F_2(y)$, $F_3(y)$, ..., and if as $n\to\infty$ so $f_n(x)\to f(x)$ and $F_n(y)\to F(y)$, does it then follow that $F(y)$ is the Fourier transform of $f(x)$? Unfortunately the answer is by no means an unequivocal yes, and the answer depends upon which of several forms of convergence is implied when we say that $f_n(x)\to f(x)$ as $n\to\infty$.

Whether or not an equation such as (1.1) is valid at some particular value of x turns out to depend upon the behaviour of the function $f(x)$ in the vicinity of the point x. Somewhat surprisingly many versions of the theorem yield $f(x)$ even at a point of discontinuity, whilst (perhaps more surprisingly) some common forms of the theorem can fail even at a point where $f(x)$ is continuous. Various forms of local average of $f(x)$ around the point x are important in this context, and we discuss these in chapter 5. This background then allows chapters 6–11 to be devoted to the classical theory of Fourier transforms.

In chapters 12–16 we go beyond classical theory and introduce generalized functions (distributions). The reader is probably already familiar with the idea that the Fourier transform of the function $f(x) = 1$ is the Dirac 'delta function' $\delta(y)$, and that one way of justifying this is to regard the function $f(x) = 1$ as the limit as $a \to \infty$ of $\exp(-\pi x^2/a^2)$, whilst $\delta(y)$ is represented by the limit $a \to \infty$ of $a \exp(-\pi a^2 y^2)$. The delta function is an example of a generalized function, and it is not a function in the traditional sense of the word. It is but one example of many other generalized functions which are particularly useful in Fourier theory.

Fourier series are covered in chapters 15 and 16. By leaving this topic till last, rather than considering it first as is often the case, we are able to use the concepts of generalized functions to show that the Fourier series analysis of periodic functions is simply a special case of the Fourier transformation.

What is being attempted in this book is a descriptive survey of several rigorously stated theorems in Fourier analysis, and not a systematic presentation of proofs. It is hoped, however, that the survey will act as a useful introduction to the understanding of those books where the proofs are to be found. Some books that the author has found particularly useful are listed in the bibliography.

2

Lebesgue integration

2.1 Introduction

Early developments of Fourier theory were based on the theory of integration of G. F. B. Riemann (1826–66). However, an alternative approach to integration developed by H. L. Lebesgue (1875–1941) turns out to be more powerful and simpler to use so far as Fourier theory is concerned, and virtually all modern mathematical approaches are based on Lebesgue's theory. Whilst a full understanding of Lebesgue's theory is not necessary in later chapters, some familiarity with the ideas is essential, and the following description is intended for a reader who is familiar with Riemann but not Lebesgue integration.

2.2 Riemann integration

The procedure for defining the proper Riemann integral of some single valued, real function of a real variable between the finite limits a and b may be visualized by dividing the area under the graph of the function into vertical strips. The area of each strip is approximated to the product of the width of the strip and the sampled value of the function at some arbitrary point within the strip, and one considers the limit of the sum of these approximate areas as the widths of the strips tend to zero and the number of such strips tends to infinity.

A rigorous treatment, as in Apostol (1974) for instance, considers whether the limit so obtained depends upon the particular way the strips or the sample values are chosen at each stage of the limiting process. In order that the limit shall not depend on these choices, it is necessary though not sufficient, that the limits of integration, a and b, are finite (i.e. positive, negative or zero real numbers) and that the function is bounded on (a, b). We remind the reader that the symbol (a, b), for $b>a$, represents the *open interval* corresponding to the set of values of x satisfying $a<x<b$, whilst

$[a, b]$ represents the *closed interval* consisting of the set of values satisfying $a \leqslant x \leqslant b$. The symbols $(a, b]$ and $[a, b)$ have analogous meanings. A real or complex valued function f is said to be bounded on an interval I if there exists a positive number A such that

$$|f(x)| < A$$

for all $x \in I$, i.e. for all x belonging to the set of points on the interval I.

The proper Riemann integral is thus not capable of dealing with infinite limits of integration or with infinite singularities in the function. To deal with these infinities the so-called improper Riemann integrals depend upon two further limiting processes. These are well known and are typified by the following two examples, for $\varepsilon > 0$ and $\lambda > 0$,

$$\int_0^1 |x|^{-1/2} \mathrm{d}x = \lim_{\varepsilon \to 0} \int_\varepsilon^1 |x|^{-1/2} \mathrm{d}x$$

$$\int_0^\infty (1 + x^2)^{-1} \mathrm{d}x = \lim_{\lambda \to \infty} \int_0^\lambda \frac{1}{1 + x^2} \mathrm{d}x.$$

In contrast, as we shall see, the Lebesgue integral can deal with infinite limits of integration, and with certain infinite singularities all in one single limiting process. Herein lies one of the great advantages of Lebesgue's method. Before describing the Lebesgue integral, however, we need to introduce the idea of a null set.

2.3 Null sets

A set of real numbers (or, in other words, of points on the real axis), is roughly speaking a null set if the points occupy a portion of the axis of zero total length, in the same way that a single point does. A finite set of points is a null set, as also is a denumerable infinity of points (i.e. an infinity of points that can be arranged in a one to one correspondence with the integers $1, 2, 3, \ldots$). For example, the set of all rational numbers (those that can be expressed as m/n, where m is a positive or negative integer or zero, and n is a positive integer) are known to be denumerable, and is thus a null set. However, a non-denumerable infinity of numbers may or may not be a null set. For example, the set of all numbers x satisfying $1 < x < 2$ is non-denumerable and is clearly not a null set, whilst the Cantor ternary set, also non-denumerably infinite, is a null set. The Cantor ternary set consists of those numbers x that can be expressed as

$$x = \sum_{n=1}^{\infty} a_n 3^{-n},$$

where each a_n is either 0 or 2.

In dealing with these less obvious situations a more precise definition is necessary, namely: a set of real numbers, or equivalently of points on the real axis, is a null set if, given any $\varepsilon>0$, it is possible to enclose the points in a finite or denumerable infinite set of open intervals whose total length is less than ε.

Some useful turns of phrase go with the concept of a null set, an alternative name for a null set being a *set of measure zero*. If some property holds for all real values of x except for a null set, we say the property holds *almost everywhere* (a.e.) or for *almost all* x (a.a.x). We say a property holds a.e. on some set S of real values of x if it holds for values of x belonging to the set S except for a null set. If two functions f and g are defined a.e. and are equal a.e. so that

$$f(x)=g(x) \qquad (\text{a.a.}x),$$

then we say the functions are *essentially equal*. The adjective 'essentially' is also useful in other contexts, to indicate that we exclude from consideration behaviour on a null set. For instance, a complex valued function f defined a.e. on the real line is said to be *essentially bounded* on $(-\infty, \infty)$ if there exists a positive number A such that $|f(x)|<A$ at a.a.x. In such a case there will be a least possible choice for A, and this minimum number is called the *essential supremum* of $|f(x)|$ on $(-\infty, \infty)$. Note that the empty set is a null set, so that if some property holds for all x then it also holds at a.a.x.

The idea of a null set can be extended in a natural manner to apply to sets of points in spaces of two or more dimensions, by requiring that the points can be enclosed in regions of arbitrarily small area, volume, etc.

2.4 The Lebesgue integral

We now describe an approach to the Lebesgue theory, chosen from several alternatives, which is based on sequences of step functions. A *step function* is here taken to be a real valued function defined everywhere on the real line such that (i) $f(x)$ is equal to a non-negative constant on each of a finite number of intervals $[a_1, a_2)$, $[a_2, a_3)$, $[a_3, a_4), \ldots, [a_{n-1}, a_n)$, where

$$a_1<a_2<a_3<\cdots<a_{n-1}<a_n,$$

the constants being not necessarily equal, and (ii) $f(x)$ is elsewhere equal to zero.

We now consider an infinite sequence of such step functions, $f_1(x)$, $f_2(x)$, $f_3(x), \ldots$, the sequence being represented by the symbol $\{f_n\}_{n=1}^{\infty}$ for convenience, and we call it an *increasing sequence of step functions* if, for every value of x and each n,

$$f_{n+1}(x)\geqslant f_n(x).$$

There are no other restrictions on the behaviour of an increasing sequence of step functions as $n \to \infty$; the heights of the flat portions may tend to infinity, the number of steps may tend to infinity, the length of the portion of the real axis covered by the steps may tend to infinity, the individual widths of the steps may tend to zero, and the position of the steps in consecutive members of the sequence need not be related. The integral I_n of each step function f_n is defined in the obvious way as the sum of the products of width and height for each flat portion in the graph of $f_n(x)$.

Now consider some real valued function f, defined a.e. on the real line, with $f(x)$ nowhere negative. If there exists an increasing sequence of step functions $\{f_n\}_{n=1}^{\infty}$ such that

$$f(x) = \lim_{n \to \infty} f_n(x) \qquad \text{(a.a.}x\text{)},$$

and, if the integrals I_n of the step functions tend to some finite number I, then we define the Lebesgue integral of f as equal to I and use the following equivalent symbols to represent the integral:

$$\int f = \int_{-\infty}^{+\infty} f(x)\mathrm{d}x = \int f(x)\mathrm{d}x = I. \tag{2.1}$$

It can be shown that, if this limit exists, it will *not* depend upon the particular sequence of step functions used. It may be useful to think of the convergence of the f_n towards f in terms of the graphs of the f_n, pushing forever upwards and outwards underneath the graph of f, gradually filling up all available gaps.

The treatment is now completed in the following ways to cover wider classes of functions. The integral of the difference of two non-negative functions f and g is *defined* as

$$\int (f-g) = \int f - \int g. \tag{2.2}$$

If $(f-g)$ is itself directly integrable by means of an increasing sequence of step functions, this definition leads to no contradictions, but more importantly the definition increases the range of functions that can be integrated. In an obvious way it allows us to deal with functions which have negative as well as positive function values; one integrates the positive going and negative going portion separately (after inverting the latter) and then subtracts the two integrals appropriately. In a much less obvious way, Weir (1973), (2.2) even allows certain positive functions to be integrated, when a direct treatment via step functions fails. The definite integral, written $\int_a^b f$ or as $\int_a^b f(x)\mathrm{d}x$, for $b \geqslant a$, is defined by assigning to $f(x)$ the value zero for values of x not on the interval $[a, b]$, and then evaluating the integral over all x. For $b < a$ we define $\int_a^b f$ as equal to the negative of $\int_b^a f$. The semi-infinite integrals $\int_a^{\infty} f$ and $\int_{-\infty}^{a} f$ are dealt with similarly.

Complex valued functions are treated by integrating the real and imaginary parts separately.

We now see how it comes about that the Lebesgue approach can deal with infinite singularities and with infinite limits of integration all in one limiting process, in contrast to the case with Riemann integration. This fact leads to certain subtleties in nomenclature which we now describe.

2.5 Nomenclature

In the Riemann theory the following symbols are equivalent by definition:

$$\int_a^\infty f(x)\mathrm{d}x, \qquad \lim_{b\to\infty}\int_a^b f(x)\mathrm{d}x.$$

In the Lebesgue theory, however, they have different meanings. The left hand symbol represents the Lebesgue integral of f over the interval (a, ∞) obtained *directly* by using sequences of step functions, whilst the right hand symbol represents the limit as b tends to infinity of a Lebesgue integral over (a, b), which amounts to two limiting processes carried out sequentially. If both symbols have meaning, then indeed they will have the same value, but there are occasions when $\lim_{b\to\infty}\int_a^b f$ will exist whilst $\int_a^\infty f$ will not. The commonest example, met frequently in Fourier theory, is the function $\sin x/x$, defined to have unit value at $x=0$. The Lebesgue integral

$$\int_0^\infty \frac{\sin x}{x}\mathrm{d}x$$

does not exist, since the positive and negative portions each have 'infinite area' due to slow decay of the function at infinity. Nevertheless, for $0<a<\infty$, $\int_0^a (\sin x/x)\mathrm{d}x$ does exist and moreover

$$\lim_{a\to\infty}\int_0^a \frac{\sin x}{x}\mathrm{d}x=\frac{\pi}{2}.$$

We may think of this limiting process as allowing the infinite areas above and below the x-axis to be cancelled out in a controlled fashion.

Similar considerations can apply around a singularity. The Lebesgue integral

$$\int_0^\infty \frac{\sin(1/x)}{x}\mathrm{d}x$$

does not exist, on account of oscillations in the value of the integrand which become infinitely large and rapid as x tends to zero, the 'areas above and below the axis being infinite'. However, it is nevertheless true that, for

$0<a<\infty$,

$$\lim_{a\to 0}\int_a^\infty \frac{\sin(1/x)}{x}\,dx=\frac{\pi}{2}.$$

The class of functions that are integrable on $(-\infty,\infty)$ by the Lebesgue technique is denoted by L. We write $f\in L$, or f is L, or $\int f<\infty$, when f is a real or complex valued function defined a.e. on the real line and when the Lebesgue integral $\int f$ exists as a finite number. We write $f\in L(a,b)$ when f is a real or complex valued function defined a.e. on the interval (a,b) and when the Lebesgue integral $\int_a^b$ exists as a finite number. $L(a,\infty)$ and $L(-\infty,a)$ are used in an analogous and obvious way. It is immaterial whether we use open or closed intervals, (a,b) or $[a,b]$, in this context since the Lebesgue integral of a function is not altered if the value of a function is changed on a null set of values of x. Indeed a function may be undefined on a null set, and still be Lebesgue integrable.

One final point of terminology. Some authors regard a positive valued function f as Lebesgue integrable, and write,

$$\int f=\infty$$

when $\lim_{n\to\infty} I_n$ considered in section 2.4 tends to infinity. The value ∞ is regarded in this context as a legitimate value for a Lebesgue integral. A function such as $\sin x/x$ is still, however, regarded as non-integrable because the positive going and negative going portions both give an infinite integral and meaning is not given to $(\infty-\infty)$. Although this alternative terminology has certain advantages we will not adopt it.

2.6 Conditions for integrability; measurability

The following well-known theorem due to Lebesgue gives sufficient and necessary conditions for a function to possess a proper Riemann integral on an interval $[a,b]$, see Apostol (1974).

Theorem 2.1 A real or complex valued function f defined everywhere on $[a,b]$ will possess a proper Riemann integral on $[a,b]$ if, and only if, f is both bounded on $[a,b]$ and also continuous a.e. on $[a,b]$. □

It can also be shown that this condition on f is sufficient, but not necessary, to ensure that $f\in L(a,b)$, from which it follows that if one or other type of improper Riemann integral exists then so also will the corresponding limit of Lebesgue integrals.

The Lebesgue method of integration is thus more powerful than that of Riemann, the extra class of functions which it can deal with being highly pathological in having discontinuities at points which are not confined to a null set. For instance, the function g defined to have $g(x)=0$ when x is rational and $g(x)=1$ when x is irrational (i.e. not rational), is nowhere continuous, and so is not Riemann integrable on $[0, 1]$; it is $L(0, 1)$, however, with $\int_0^1 g=1$. We remind the reader that a function is defined as continuous at x if, and only if,

$$f(x^+)=f(x)=f(x^-),$$

where $f(x^+)$ and $f(x^-)$ are the right and left hand values defined, for $\varepsilon>0$, by

$$f(x^+)=\lim_{\varepsilon\to 0} f(x+\varepsilon)$$

$$f(x^-)=\lim_{\varepsilon\to 0} f(x-\varepsilon).$$

The function $g(x)$ defined above does not possess right or left hand values at any value of x. A function $f(x)$ is said to be discontinuous at x if it is not continuous at x; x is then a point of discontinuity of f.

A simple necessary and sufficient condition for Lebesgue integrability does not exist. It is very useful, however, to have a feeling for a hierarchy of three conditions that a function must satisfy in order to be L. These three conditions concern, respectively, measurability, local behaviour around infinite singularities, and decay at infinity. We tackle measurability, the least familiar of the three, first.

Suppose we start with some real valued function f which is defined a.e. on the real line but is not necessarily L. Suppose we now form from f another function f_T by truncating f above and below and outside of some finite interval (a, b). In other words we define f_T, for some $A>0$, by:

$$\begin{aligned}
f_T(x)&=f(x) && \text{if } a<x<b \text{ and } |f(x)|<A,\\
f_T(x)&=A && \text{if } a<x<b \text{ and } f(x)\geqslant A,\\
f_T(x)&=-A && \text{if } a<x<b \text{ and } f(x)\leqslant -A,\\
f_T(x)&=0 && \text{if } x\geqslant b \text{ or if } x\leqslant a,\\
f_T(x)&=0 && \text{if } f(x) \text{ is undefined at } x.
\end{aligned}$$

If now $f_T\in L$ for every finite (a, b) and every $A>0$ then we say that f is measurable. A complex valued function is said to be measurable if its real and imaginary parts are measurable. A function must be measurable in order to be L, but not vice versa. In 'topping and tailing' the function f to form f_T we have removed those infinities which so commonly stop a function being integrable; if the function f_T is *still* not integrable it indicates that $f(x)$ displays pathologically irregular fluctuations in value between arbitrarily close values of x. Examples of non-measurable functions are given for instance in Weir (1973): they show that a function may be defined and bounded on $[a, b]$ and yet not be $L(a, b)$.

A more stringent condition than measurability is that of local integrability. A function f is defined as *locally integrable* if it is $L(a, b)$ whenever a and b are finite, and the class of locally integrable functions is written L_{LOC}. The additional constraint implied by local integrability, over and above that implied by measurability, has to do with the nature of infinite singularities in the function. For instance, the functions equal a.e. to $1/x$ and $1/|x|^{1/2}$, respectively, are both measurable, but only the latter is locally integrable, due to better behaviour around $x=0$. It can be shown that if a measurable function is essentially bounded on $[a, b]$ then it is necessarily $L(a, b)$. Boundedness by itself, without measurability, does not, however, ensure local integrability.

Finally, the condition of integrability is more stringent than local integrability. For instance, the functions equal a.e. to $|x|^{-1/2}$ and $1/(1+x^2)$, respectively, are both locally integrable, but only the latter is L, since $|x|^{-1/2}$ decays too slowly as $x \to \pm\infty$. The extra constraint implied by integrability, over that implied by local integrability, has to do with the behaviour of the function as x tends to $\pm\infty$. However, a function that is $L(-\infty, \infty)$ does not necessarily tend to zero at infinity. For instance, consider the function $h(x)$ whose graph consists of an infinite set of rectangular pulses centred on $x = \pm 1, \pm 2, \ldots, \pm n, \ldots$, the pulses at $x = \pm n$ having height n and width n^{-3}; we find $f \in L$, but it is not true that $\lim_{x \to \pm\infty} f(x) = 0$.

Following from the concept of measurability a quantity called the measure $m(E)$, of a set of points E on the real line can be defined, acting as a generalization of the concept of length. First we define a function $g(x)$ as being the *indicator function* of the set E of points on the real line if $g(x) = 1$ when $x \in E$, and $g(x) = 0$ when $x \notin E$. If $g \in L$, then we say the set E is *measurable*, and we define $m(E)$ as equal to $\int g$; otherwise the set E is not measurable. A null set has $m(E) = 0$.

If a function $f(x)$ and its modulus $|f(x)|$ are both Riemann integrable over $(-\infty, \infty)$, employing the method of the improper integral around at most a finite number of discontinuities, then necessarily $f \in L(-\infty, \infty)$: the converse, however, does not follow. The following provides another important way of establishing Lebesgue integrability.

***Theorem* 2.2** Let f be a real or complex valued function that is locally Lebesgue integrable and suppose that, for some $g \in L(-\infty, \infty)$,

$$|f(x)| \leqslant |g(x)| \quad \text{a.e.:}$$

then it will follow that $f \in L(-\infty, \infty)$. □

Other ways of establishing Lebesgue integrability will appear in section 4.3 when dominated and monotone convergence are discussed.

2.7 Functions in L^p

A class of functions designated as L^p is of great importance in the theory of Fourier transformation.

Definition If f is a locally integrable function such that $|f|^p \in L$, where p is a (real) number satisfying $p \geqslant 1$, then we say that f is pth power Lebesgue integrable: the set of all such functions is written L^p. □

The class L^1 is clearly synonymous with the class L, since if $f \in L$ then also $|f| \in L$. Note, however, that the fact that $|f| \in L$ does not imply that $f \in L$, since the modulus operation may disguise pathological oscillations in sign of f that render f not measurable. If $f \in L^2$ then we say that f is square integrable, or quadratically integrable. Many functions in the physical sciences, such as wave amplitudes in classical or quantum situations, are square integrable, and the class L^2 is thus of particular importance.

Subsidiary definitions over restricted domains follow naturally enough. If f is a complex valued function defined a.e. on some set of points Q on the real line, then we write $f \in L^p(Q)$ if $g \in L^p$, where g is a 'completion' of f, defined so that $g(x)$ is equal to $f(x)$ a.e. on Q and to zero elsewhere on the real line. Thus in particular $L^p(a, b)$, $L^p(a, \infty)$, $L^p(-\infty, a)$ are given meaning by choosing Q to be the intervals (a, b), (a, ∞), $(-\infty, a)$, respectively. We say f is locally L^p if f is $L^p(a, b)$ for all finite values of a and b, and we then write $f \in L^p_{\mathrm{LOC}}$, or $f \in L_{\mathrm{LOC}}$ when $p = 1$.

The fact that a function is L^p for one value of p does not imply that it will be for some other value of p. For instance, $(1+|x|)^{-1}$ is L^2 but not L^1, whilst $|x|^{-1/2}\,\mathrm{e}^{-|x|}$ is L^1 but not L^2. Some statements along these lines can be made, however. If a function is locally L^p, for some $p \geqslant 1$, then it will be locally L^q for all q satisfying $1 \leqslant q \leqslant p$. In other words raising a function to some power $p > 1$ makes the infinite singularities get 'worse' as p is increased. On the other hand, if a function is essentially bounded on $(-\infty, \infty)$ and is L^p for some $p \geqslant 1$, then it will be L^q for all $q \geqslant p$. In other words raising an essentially bounded function to some power p makes its behaviour at infinity get 'better' so far as integrability is concerned. For instance, the function $(1+|x|)^{-1} \in L^{1 \cdot 1}$, and is also bounded on $(-\infty, \infty)$: it is thus also square integrable. If the boundedness condition is relaxed this result no longer holds, even if the function is still *locally* bounded (i.e. bounded on every finite interval). The function $h(x)$ considered in section 2.6, page 11, is bounded on every finite interval (a, b), and is L^1, but is not L^2.

Finally it is very useful to give meaning to the case $p = \infty$. The convenience of the notation will become clear when we consider theorems due to Holder and Young which are of relevance to products and convolutions. A complex valued function, defined a.e. on the real line, is defined as belonging to class L^{∞} if it is measurable and essentially bounded on $(-\infty, \infty)$. So far as this definition is concerned we could replace 'measurable' by 'locally integrable', because of the boundedness. In section 3.1 we will quote a result showing how the case $p=\infty$ can be regarded as the limiting case as $p \to \infty$.

2.8 Integrals in several dimensions

The construction of a Lebesgue integral in one dimension, described in section 2.4, is readily extended to integrals in several dimensions. For instance, in two dimensions use is made of an increasing sequence of step functions, $\{f_n(x, y)\}_{n=1}^{\infty}$, which tends a.e. to the limit function $f(x, y)$. The graph of such a step function can be visualized as a finite set of rectangular blocks whose bases lie on the xy-plane. The integral of such a step function is calculated from the volumes of these blocks, and the rest of the treatment is closely analogous to the one-dimensional case. The final Lebesgue integral of a function $f(x, y)$ over the whole of two-dimensional space is written in either of the following ways:

$$\int f, \qquad \int f(x, y)\mathrm{d}(x, y).$$

A definite integral over some region Q is defined by setting $f(x, y)$ to zero for points not belonging to Q, and then integrating over the whole plane: the result is written as

$$\int_Q f \qquad \text{or as} \qquad \int_Q f(x, y)\mathrm{d}(x, y).$$

Note that these multidimensional integrals still rely on only one limiting process so that it is appropriate to use only one integral sign, $\int$, no matter how many dimensions are involved. Such *single* integrals in several dimensions are to be clearly distinguished from the following *repeated* integrals, which consist of sequences of integrals carried out one after the other. Suppose that $\int f(x, y)\mathrm{d}x$ exists for a.a.y, and that the resulting function (of the variable y) is L; we now define the repeated integral $\int\int f(x, y)\mathrm{d}x\,\mathrm{d}y$ by

$$\int\int f(x, y)\mathrm{d}x\,\mathrm{d}y = \int\left[\int f(x, y)\mathrm{d}x\right]\mathrm{d}y.$$

This is, in turn, distinct from the following repeated integral in which the order of x- and y-integration is the opposite of that above:

$$\int\int f(x, y)\mathrm{d}y\,\mathrm{d}x = \int\left[\int f(x, y)\mathrm{d}y\right]\mathrm{d}x.$$

The theorems of Fubini and of Tonelli, see chapter 3, give circumstances when these various single and repeated integrals are equal. When we say a function f is Lebesgue integrable in n dimensions we mean that the single integral exists, and we write $f \in L$. If there is any ambiguity about the dimensionality of the domain over which the integration applies we write $f \in L(R^n)$ for integrability over the whole of n-dimensional Euclidean space, or $f \in L(Q)$ for integrability over some specified region Q of such a space.

The concept of measurability is adaptable also, in an analogous fashion, to spaces of several dimensions. Likewise the class $L^p(\mathbf{R}^n)$, for $p \geqslant 1$, are readily given meaning, along with $L^p(Q)$, referring to domains within n-dimensional Euclidian space.

2.9 Alternative approaches

The approach described in this chapter is commonly used nowadays and is described in detail, for instance, in Weir (1973); however, it differs from the approach originally used by Lebesgue, although the final results are equivalent. In the original method, summarized for instance in Titchmarsh (1962) and Wiener (1951), one starts by constructing a definition of the measure $m(E)$ of a set of points E. The definition of a measurable function then follows, and this leads finally to the definition of the Lebesgue integral.

3

Some useful theorems

3.1 The Minkowski inequality

In this chapter we describe some theorems which are used frequently in the background to Fourier theory; we start with a result concerning the sum of two functions.

For two complex numbers A and B it is clear that $|A \pm B| \leqslant |A| + |B|$. Much the same idea is embodied in the following inequalities which hold for a pair of functions $f \in L$ and $g \in L$,

$$\left|\int f\right| \leqslant \int |f|$$

$$\int |f+g| \leqslant \int |f| + \int |g|.$$

The latter inequality is generalized to functions in L^p in an inequality named after H. Minkowski (1864–1909). It is convenient, before stating this theorem, to define the *p-norm* (or simply the norm) of a function in L^p. If $f \in L^p$, for some p satisfying $1 \leqslant p < \infty$, then we define the p-norm, $\|f\|_p$, of f as

$$\|f\|_p = \left(\int |f|^p\right)^{1/p}.$$

The case when $f \in L^\infty$ requires a separate definition, and given a function $f \in L^\infty$ we define $\|f\|_\infty$ as equal to the essential supremum of $|f|$. This choice of definition for $\|f\|_\infty$ receives a justification in the following.

Theorem 3.1 Consider a function $f \in L^\infty$ that also belongs to L^p for some p satisfying $1 \leqslant p < \infty$: then it will follow that $f \in L^r$ whenever $p < r < \infty$ and that

$$\lim_{r \to \infty} \|f\|_r = \|f\|_\infty. \quad \square$$

The Minkowski theorem may now be stated.

***Theorem* 3.2** Suppose $f \in L^p$ and $g \in L^p$ for some p satisfying $1 \leqslant p \leqslant \infty$: then it follows that

$$\|f+g\|_p \leqslant \|f\|_p + \|g\|_p. \quad \square$$

This inequality is an example of a class of so-called *triangle inequalities*, the name coming from the analogy with the geometric inequality relating the lengths of the sides of a triangle.

3.2 Hölder's theorem

We now describe a theorem named after O. Hölder (1859–1937) which relates to the product of two functions. Whilst if $f \in L$ and $g \in L$ it follows that $(f \pm g) \in L$, it does not follow that the product $fg \in L$. This failure may either be due to the superposition of local infinite singularities (consider the product of the integrable function $|x|^{-1/2} e^{-|x|}$ with itself), or it may be due to poor behaviour of the product at infinity (consider the product of the integrable function $h(x)$, described in section 2.6, page 11, with itself). It can be shown, however, that if $f \in L$ and $g \in L^\infty$, then $fg \in L$; additionally it can be shown also that if f and g are both L^2 then $fg \in L$. These two cases are special instances of a more general result due to Hölder which may be stated as follows.

***Theorem* 3.3** Suppose $f \in L^p$ and $g \in L^r$, where p and r satisfy $1 \leqslant p \leqslant \infty$, $1 \leqslant r \leqslant \infty$, $0 \leqslant p^{-1} + r^{-1} \leqslant 1$: then it will follow that $fg \in L^P$, where $P^{-1} = p^{-1} + r^{-1}$, and that

$$\|fg\|_P \leqslant \|f\|_p \|g\|_r,$$

it being understood throughout that ∞^{-1} and zero are interchangeable in the expressions involving p^{-1}, r^{-1}, or P^{-1}. $\square$

The well-known *Schwarz inequality* for square integrable functions corresponds to the special case $p = r = 2$, and may be written more explicitly as

$$\left(\int |fg|\right)^2 \leqslant \left(\int |f|^2\right)\left(\int |g|^2\right),$$

or since, for any $h \in L$, $|\int h| \leqslant \int |h|$, as

$$\left|\int fg\right|^2 \leqslant \left(\int |f|^2\right)\left(\int |g|^2\right). \tag{3.1}$$

This inequality lies behind the uncertainty principle of quantum mechanics, or the equivalent band-width theorem of communication theory.

As examples of the use of Holder's theorem, we can readily use it to show that if $f \in L^p$, for some p satisfying $1 < p < \infty$, then $f(x)/(1+|x|) \in L$.

Alternatively if $f \in L^\infty$, then $f(x)/(1+|x|)^s \in L$, where s is any real number greater than unity. These examples show how the function $(1+|x|)^{-s}$ can be used to modify the behaviour at infinity of some function in L^p so as to bring it into class L.

3.3 Young's theorem

This theorem due to W. H. Young (1863–1942) bears somewhat the same relation to a convolution product as does Hölder's theorem to an ordinary product. Let us first define a convolution product. Suppose f and g are real or complex valued functions defined a.e. on the real axis; the convolution of f and g, written $f * g$, is defined as a function having the value $f * g(x)$ at any value of x for which the following integral exists:

$$f * g(x) = \int f(x')g(x-x')\mathrm{d}x'. \tag{3.2}$$

It is readily shown that $f * g$ is the same as $g * f$.

We are usually interested in cases when $f * g$ is defined a.e., and it is clearly useful to know sets of conditions on f and g that will ensure that this is so. Two commonly met cases are as follows. If $f \in L$ and $g \in L$ then $f * g \in L$, whilst alternatively if f and g are both square integrable then $f * g$ is everywhere continuous and is in L^∞. Young's theorem is more general* and may be stated as follows, using the usual convention that ∞^{-1} and zero are interchangeable.

Theorem 3.4 Suppose $f \in L^p$ and $g \in L^r$, where p and r satisfy $1 \leqslant p \leqslant \infty$, $1 \leqslant r \leqslant \infty$, $1 \leqslant p^{-1} + r^{-1} \leqslant 2$: then it will follow that (i) $f * g \in L^Q$, where Q is given by $Q^{-1} = p^{-1} + r^{-1} - 1$, (ii) whenever $p^{-1} + r^{-1} = 1$ then $f * g$ is everywhere continuous and is bounded on $(-\infty, \infty)$, and (iii)

$$\|f * g\|_Q \leqslant \|f\|_p \|g\|_r. \quad \square$$

Young's theorem can be applied repetitively to deal with convolutions of three or more functions, and, for example, with three functions $f \in L^p, g \in L^r$, $h \in L^u$ we find

$$\|f * g * h\|_Q \leqslant \|f\|_p \|g\|_r \|h\|_u,$$

where each of p, r, u, Q has a value on $[1, \infty]$ and where $p^{-1} + r^{-1} + u^{-1} = 1 + Q^{-1}$. The triple convolution is the same function irrespective of the ordering or bracketing of the functions.

The *cross correlation*, ρ_{fg}, of two functions f and g is defined as follows,

* For a tighter version of the inequality in theorem 3.4 see Beckner (1975).

at each value of x for which the integral converges,

$$\rho_{fg}(x)=\int f^*(x')g(x+x')\mathrm{d}x', \tag{3.3}$$

where the asterisk indicates complex conjugation. In Young's theorem the convolution $f*g$ can be replaced by the correlation ρ_{fg}. Indeed a comparison of (3.2) with (3.3) shows that

$$\rho_{fg}=f_{\mathrm{T}}^* * g \tag{3.4}$$

where f_{T}^* is the complex conjugate of the transpose of f. From this it follows that the order of the functions in a cross correlation does matter, and ρ_{fg} is not necessarily equal to ρ_{gf}: indeed at each x for which one or other quantity is defined we have

$$\rho_{fg}(x)=[\rho_{gf}(-x)]^*. \tag{3.5}$$

Throughout Fourier theory, results on convolution have their counterparts phrased in terms of correlations; we will concentrate on the former.

3.4 The Fubini and Tonelli theorems

Repeated integrals occur frequently in Fourier theory, the Fourier integral of a convolution product being, for instance, a repeated integral, and it is often useful to be able to invert the order of the integrations. The following theorem due to G. Fubini (1879–1943) is sometimes useful in this respect.

Theorem 3.5 If a real or complex valued function $f(x, y)$ is integrable over the xy-plane then the following integrals all exist and are equal:

$$\int f(x, y)\mathrm{d}(x, y)=\int\int f(x, y)\mathrm{d}x\,\mathrm{d}y=\int\int f(x, y)\mathrm{d}y\,\mathrm{d}x. \quad \square \tag{3.6}$$

Note, however, that unfortunately the existence of one of the repeated integrals is not of itself sufficient to ensure the existence of either of the other integrals. The following extension of the theorem, due to Tonelli, gives a sufficient condition for this.

Theorem 3.6 Let $f(x, y)$ be a real or complex valued function that is defined a.e. on the xy-plane and is also measurable, and suppose that at least one of the repeated integrals $\int\int|f(x, y)|\mathrm{d}x\,\mathrm{d}y$, $\int\int|f(x, y)|\mathrm{d}y\,\mathrm{d}x$ exists: then it will follow that f is integrable over the xy-plane so that the conditions of theorem 3.5 are satisfied and (3.6) will be valid. $\square$

Note importantly that in the conditions of theorem 3.6 it is the modulus, $|f|$, rather than f itself which appears in the integrands of the repeated integral. The measurability of $f(x, y)$ can be guaranteed in certain commonly met situations. For instance, $f(x, y)$ will be measurable if f is integrable over each finite rectangular area of the xy-plane. Also, if f and g are locally integrable functions of a single real variable, then the functions $f(x+y)$, $f(x)g(y)$ and $f(x)g(x+y)$ will be measurable on the xy-plane.

The Fubini–Tonelli theorems have natural counterparts in spaces of higher dimensions, the repeated integrals in this case covering spaces of dimensionality lower than that of the function itself. For example, if $f(x, y, z)$ is a measurable function in three dimensions, then the existence of one or other of

$$\int f(x, y, z)\mathrm{d}(x, y, z)$$

or of the thrice repeated integral

$$\int\int\int |f(x, y, z)|\mathrm{d}x\,\mathrm{d}y\,\mathrm{d}z$$

or of the twice repeated integral

$$\int\int |f(x, y, z)|\mathrm{d}x\,\mathrm{d}(y, z),$$

is sufficient to ensure that

$$\begin{aligned}\int f(x, y, z)\mathrm{d}(x, y, z) &= \int\int\int f(x, y, z)\mathrm{d}x\,\mathrm{d}y\,\mathrm{d}z \\ &= \int\int f(x, y, z)\mathrm{d}x\,\mathrm{d}(y, z) \\ &= \int\int\int f(x, y, z)\mathrm{d}y\,\mathrm{d}x\,\mathrm{d}z \\ &= \int\int f(x, y, z)\mathrm{d}(y, z)\,\mathrm{d}x\end{aligned}$$

etc.

3.5 Two theorems of Lebesgue

Although a function f can be discontinuous everywhere and yet still remain Lebesgue integrable, integrability does impose some sort of constraint on the fluctuations in value of $f(x)$ when x is altered by a small amount. The following theorems exemplify this in different ways.

***Theorem** 3.7* Given a function $f \in L^p$ for some p satisfying $1 \leqslant p < \infty$, let f_δ be the function defined a.e. by $f_\delta(x) = f(x+\delta)$, where δ is a non-zero real number: then it follows that $f_\delta \in L^p$ and that

$$\lim_{\delta \to 0} \| f_\delta - f \|_p = 0. \quad \square$$

When theorem 3.7 is combined with Hölder's theorem 3.3, the following rather more general result follows.

Theorem 3.8 Suppose $f \in L^p$ and $g \in L^r$, where p and r satisfy $1 \leqslant p < \infty$, $1 \leqslant r \leqslant \infty$, $0 < p^{-1} + r^{-1} \leqslant 1$, and ∞^{-1} and zero are regarded as interchangeable: then it follows that

$$\lim_{\delta \to 0} \| g(f_\delta - f) \|_P = 0$$

where f_δ is the function defined a.e. by $f_\delta(x) = f(x+\delta)$, and $P^{-1} = p^{-1} + r^{-1}$. $\square$

The next theorem relates to the local behaviour of an integrable function and lies behind the concept of local averaging which is important in many versions of the Fourier inversion theorem.

Theorem 3.9 Suppose f is a locally integrable function defined a.e. on the real line: then at a.a.x the following limits are both zero:

$$\lim_{\varepsilon \to 0^+} \frac{1}{\varepsilon} \int_0^\varepsilon |f(x+u) - f(x)| \mathrm{d}u = 0 \tag{3.7}$$

$$\lim_{\varepsilon \to 0^+} \frac{1}{\varepsilon} \int_0^\varepsilon |f(x-u) - f(x)| \mathrm{d}u = 0. \quad \square \tag{3.8}$$

The set of points at which (3.7) and (3.8) are valid is often called the *Lebesgue set* of the function f. We return to a fuller discussion of this theorem, with examples, in section 5.2.

3.6 Absolute and uniform continuity

Fourier theorems involving differentiation rely heavily on the conditions under which differentiation and integration can be regarded as inverses. In this section we describe such conditions, and start by revising the notions of derivative and primitive.

The *differential coefficient*, $\mathrm{d}f/\mathrm{d}x$, of a function $f(x)$ is defined at a point x if, and only if, the following two limits both exist and are equal:

$$\lim_{\delta \to 0^+} \left\{ \frac{f(x+\delta) - f(x)}{\delta} \right\} = \lim_{\delta \to 0^+} \left\{ \frac{f(x-\delta) - f(x)}{\delta} \right\};$$

this limit is then written as $\mathrm{d}f/\mathrm{d}x$. The function defined as equal to $\mathrm{d}f/\mathrm{d}x$ at each point where the latter is defined is called the *derivative* of f, and is usually written f'.

If a function g is $L(a, b)$, then, for all x on the finite interval $[a, b]$, we may define a function $f(x)$ by

$$f(x)=\int_a^x g(u)\mathrm{d}u+C, \tag{3.9}$$

where C is an arbitrary constant. This function f will be continuous on (a, b), will have $f(a)=f(a^+)$ and $f(b)=f(b^-)$, and f' will be defined and equal to g a.e. on $[a, b]$ so that $g(u)$ may be replaced by $f'(u)$ in (3.9). Any such function f, regardless of the value of C, is called an *indefinite integral (primitive)* of g on $[a, b]$. If $g \in L_{\mathrm{LOC}}$ and (3.9) holds for all $x \in (-\infty, \infty)$, for fixed and arbitrary a and C, then we say that f is a primitive of g on $(-\infty, \infty)$.

The fact that some function, say f, is continuous and differentiable on an interval does not, however, ensure that it is the primitive of its derivative, i.e. that

$$f(x)=f(a)+\int_a^x f'(u)\mathrm{d}u. \tag{3.10}$$

Consider, for example,

$$f(x)=\begin{cases} x^2\cos(1/x^2), & x\neq 0 \\ 0, & x=0 \end{cases}.$$

The derivative f' is defined *everywhere* (note $f'(0)=0$), yet (3.10) fails on choosing $a=-1$ and $x=+1$, because f' is not $L(-1, 1)$: this arises from a divergence in $f'(x)$ as $x\to 0$. Another surprising counterexample will be described in section 11.5; this consists of a function that is continuous and monotonically increasing on $[0, 1]$ and yet has a derivative f' equal a.e. to zero on $[0, 1]$. Clearly (3.10) fails in this case also.

Absolute continuity provides a condition that is necessary and sufficient to ensure that a function is the integral of its derivative; it is defined as follows.

Definition A function f, defined on a finite closed interval $[a, b]$ is said to be *absolutely continuous* on $[a, b]$ if for each $\varepsilon>0$ there exists a $\delta>0$ such that

$$\left|\sum_{n=1}^{N}\{f(b_n)-f(a_n)\}\right|<\varepsilon$$

whenever

$$\sum_{n=1}^{N}(b_n-a_n)<\delta,$$

where the a_n and b_n are any real numbers such that

$$a\leqslant a_1<b_1<a_2<b_2<\cdots<b_N\leqslant b. \quad \square$$

When f is absolutely continuous on every finite interval $[a, b]$ we say it is absolutely continuous on $(-\infty, \infty)$.

The importance of absolute continuity stems from the following.

Theorem 3.10 A function f will be absolutely continuous on $[a, b]$ if, and only if, it is the indefinite integral over $[a, b]$ of some function, say g, that is $L(a, b)$: when this condition is satisfied it will follow that (i) the derivative f' is defined a.e. on $[a, b]$, (ii) $f' = g$ a.e. on $[a, b]$, and (iii) at each x on $[a, b]$

$$f(x) = f(a) + \int_a^x f'(u)du. \quad \square$$

The conditions which define absolute continuity are not easy to apply directly, and in practice various simpler conditions are useful which are sufficient but not necessary in ensuring absolute continuity. The following theorem valid for proper Riemann integration is often called the 'fundamental theorem of calculus'.

Theorem 3.11 Let f be a function such that (i) f and f' are defined everywhere on a finite interval (a, b), (ii) $f(a) = f(a^+)$ and $f(b) = f(b^-)$, and (iii) the proper Riemann integral of f' on $[a, b]$ exists: then for all $x \in [a, b]$ (3.10) will be valid as a Riemann integral. $\square$

The following is suitable for many applications.

Theorem 3.12 Given a function f and a finite interval $[a, b]$ suppose that (i) f is continuous on (a, b) and that $f(a) = f(a^+)$ and $f(b) = f(b^-)$, (ii) the derivative f' is defined and continuous at all except a finite number of points on $[a, b]$, and either (iii*a*) f' is bounded on the set of points in $[a, b]$ for which it is defined, or (iii*b*) $f' \in L(a, b)$: then it follows that f is absolutely continuous on $[a, b]$ and (3.10) is valid for $x \in [a, b]$. $\square$

As examples of the use of theorem 3.12 note that the functions $\exp(-|x|)$, and $|x|^{1/2}$ are absolutely continuous on $[-1, 1]$ despite the discontinuities in f' at $x = 0$. The functions $x \ln|x|$ and $x^2 \cos(1/x)$ will be absolutely continuous on $[-1, 1]$ if they are given the value zero at $x = 0$.

Absolute continuity lies behind the formula for integration by parts, as follows.

Theorem 3.13 Let f and g be absolutely continuous functions on the finite interval $[a, b]$ and let f' and g' be their derivatives: then

$$\int_a^b fg' = f(b)g(b) - f(a)g(a) - \int_a^b f'g. \quad \square$$

Uniform continuity is another form of continuity defined as follows.

Definition Suppose that f is a function defined on a set of points S on the real line: we say f is uniformly continuous on S if, for each $\varepsilon>0$ there is a $\delta>0$, depending only on ε, such that $|f(b)-f(a)|<\varepsilon$ whenever $a\in S$, $b\in S$ and $|b-a|<\delta$. □

The set S is often chosen as a finite or infinite, closed or open interval, but it need not be. Uniform continuity is often discussed in the context of uniform convergence, to be discussed in section 4.4.

Ordinary continuity of a function at every point on an open interval (a, b) does not imply uniform continuity on (a, b); for instance, the function $1/x$ is continuous but not uniformly continuous on $(0, 1)$. However, a function f will be uniformly continuous on a finite interval $[a, b]$ if, and only if, f is continuous on (a, b) and $f(a)=f(a^+)$ and $f(b)=f(b^-)$. Thus if f is continuous on (a, b) or on $[a, b]$ it will be uniformly continuous on every interval, open or closed, with end points c, d satisfying $a<c<d<b$.

Continuity on (a, b), uniform continuity on $[a, b]$, and absolute continuity on $[a, b]$ form a hierarchy of conditions, each more stringent than its predecessor.

3.7 The Riemann–Lebesgue theorem

This result, a generalization of a result obtained first by Riemann in the restricted context of Riemann integration, is as follows.

Theorem 3.14 Suppose $f\in L$ and let a be a positive number: then $f(x)\cos ax$ and $f(x)\sin ax$ are both in L and

$$\lim_{a\to\infty}\int_{-\infty}^{+\infty} f(x)\cos ax\,\mathrm{d}x=\lim_{a\to\infty}\int_{-\infty}^{+\infty} f(x)\sin ax\,\mathrm{d}x=0.\quad\square$$

The result can be made intuitively plausible by visualizing f as a continuous function changing value slowly in comparison with the oscillations in the sine or cosine functions. The areas of neighbouring oscillations in $f(x)\cos ax$ then cancel to zero ever more accurately as $a\to\infty$.

Theorem 3.14 has an immediate application in Fourier theory. If $f\in L$, then at all real y one may define a function F by

$$F(y)=\int f(x)\,\mathrm{e}^{-\mathrm{i}xy}\,\mathrm{d}y$$

and it will follow that $F(y)$ tends to zero as $y\to\pm\infty$.

4

Convergence of sequences of functions

4.1 Introduction

If a sequence of functions f_n tends to a function f as $n \to \infty$, and if the Fourier transforms F_n tend towards a function F as $n \to \infty$, does it follow that F is the Fourier transform of f? Again if $f_n \to f$ as $n \to \infty$, can we say that $\lim_{n\to\infty} \int f_n = \int f$? These questions are badly posed as they stand because the type of convergence intended has not been specified. In this chapter we describe several forms of convergence, with an eye eventually to stating answers to the above questions.

Throughout the chapter we consider real or complex valued functions defined on all or some part of the real line. We use the symbol $\{f_n\}_{n=1}^{\infty}$ or $\{f_n\}$ to represent a sequence of such functions f_n, $n = 1, 2, 3, \ldots, \infty$.

4.2 Pointwise convergence

This is the simplest and most obvious form of convergence. We say the sequence of functions $\{f_n\}_{n=1}^{\infty}$ converges pointwise (p.w.) to some function f as $n \to \infty$, and we write $\lim_{n\to\infty} f_n = f$ p.w. if at each value of x:

$$\lim_{n\to\infty} f_n(x) = f(x). \tag{4.1}$$

Alternatively we say that the sequence converges pointwise almost everywhere (p.w.a.e.) to f as $n \to \infty$, if (4.1) is valid for a.a.x. Further variants are possible depending upon the set of points at which convergence is required. One can define pointwise convergence on some set S of points on the real line, or p.w.a.e. on some set S, in obvious ways.

Pointwise convergence is not a very powerful form of convergence, since properties of the f_n such as integrability, differentiability, or continuity are not necessarily passed on to the limit function, and it does not follow that $\lim_{n\to\infty} \int f_n = \int f$. Consider, for example, the functions f_n defined by

$$f_n(x)=\begin{cases} n, & x\in(0,1/n) \\ 0, & x\notin(0,1/n) \end{cases}. \tag{4.2}$$

Although $\lim_{n\to\infty} f_n(x)=0$ pointwise at *all* x, we nevertheless have

$$\lim_{n\to\infty}\int f_n=1.$$

However, if pointwise convergence is backed up by certain other conditions then various properties can be passed on from the sequence to the limit. As examples of this we now consider bounded, dominated, monotone and uniform convergence.

4.3 Bounded, dominated and monotone convergence

The sequence $\{f_n\}$ defined by (4.2) is unbounded as $n\to\infty$, and it is natural to ask whether $\int f_n$ will tend to $\int f$ when such unbounded behaviour is prohibited. This is so provided we consider only a finite range of integration, a result embodied in the following theorem of *bounded convergence.*

Theorem 4.1 Consider a sequence of functions $\{f_n\}_{n=1}^{\infty}$ and a function f and suppose that $\lim_{n\to\infty} f_n=f$ p.w.a.e. on $[a,b]$, where a and b are finite; suppose also that $f_n\in L(a,b)$ for each n, and that there exists a positive number A, independent of n, such that for each n

$$|f_n(x)|<A, \quad \text{a.e. on } [a,b]: \tag{4.3}$$

then it follows that $f\in L(a,b)$ and that

$$\lim_{n\to\infty}\int_a^b f_n=\int_a^b f. \quad \square \tag{4.4}$$

Note particularly that there is no need to assume that $f\in L(a,b)$ from the outset, and note that this theorem provides one way of establishing integrability of a function over a finite interval.

Theorem 4.1 will fail if the range of integration is made infinite. This is demonstrated by the sequence $\{f_n\}_{n=1}^{\infty}$, where

$$f_n(x)=\begin{cases} 0, & |x|\geqslant n \\ 1/n, & |x|<n \end{cases}. \tag{4.5}$$

Even though the sequence is bounded on $(-\infty,\infty)$, (4.3), and also $\lim_{n\to\infty} f_n=0$ p.w. at all x, nevertheless

$$\lim_{n\to\infty}\int f_n=2.$$

We may, however, extend the limit of integration to infinity and even allow unbounded sequences if we use the following more general theorem of *dominated convergence.*

Theorem 4.2 Consider a sequence $\{f_n\}_{n=1}^{\infty}$ of functions and a function f and suppose $\lim_{n\to\infty} f_n = f$ p.w.a.e.; suppose also that $f_n \in L$ for each n, and that there exists a function $g \in L$, independent of n, such that for each n

$$|f_n(x)| < |g(x)| \quad \text{a.e.:}$$

then it follows that $f \in L$ and that

$$\lim_{n\to\infty} \int f_n = \int f. \quad \square$$

Note once again that integrability of f need not be assumed at the outset. Bounded convergence is readily seen to be a special case of dominated convergence, if we choose functions that are zero outside of some finite interval $[a, b]$.

A related theorem, in which even the existence of a p.w. limit function need not be assumed at the outset, is the following *monotone convergence* theorem.

Theorem 4.3 Consider a sequence $\{f_n\}_{n=1}^{\infty}$ of non-negative functions with $f_n \in L$ for all n and suppose there exists a positive number A, independent of n, such that for all n

$$\int f_n < A;$$

suppose also that the sequence is monotone in the sense that for each n

$$f_n(x) \leqslant f_{n+1}(x) \quad \text{a.e.:}$$

then there will exist a function $f \in L$ such that $\lim_{n\to\infty} f_n = f$ p.w.a.e. and

$$\lim_{n\to\infty} \int f_n = \int f. \quad \square \tag{4.6}$$

The bounded, dominated and monotone types of convergence pass integral properties on to the limit function, (4.6), but fail with local properties such as continuity and differentiability. We now turn for help in this direction to uniform convergence.

4.4 Uniform convergence

Even though a sequence $\{f_n\}$ may converge p.w. to some limit function f, it is possible that the rate of convergence may vary from point to point so that even for very large values of n there may still be certain points at which $f(x) - f_n(x)$ is not 'small'. This can apply locally around some fixed point, or at infinity. Consider, for instance, the following sequence:

$$f_n(x) = \begin{cases} 0, & x < 0 \\ 1 - e^{-nx}, & x \geqslant 0' \end{cases}$$

and define the Heaviside step function $H(x)$ by

$$H(x)=\begin{cases}0, & x\leqslant 0\\ 1, & x>0\end{cases}. \tag{4.7}$$

We find that $\lim_{n\to\infty} f_n(x)=H(x)$ p.w. at *all* x; nevertheless, no matter how large we make n, there still exist points (near $x=0$) at which $[H(x)-f_n(x)]>\frac{1}{2}$. As an example of a different kind, consider a sequence based on a rectangular pulse that moves out towards infinity:

$$f_n(x)=\begin{cases}1, & x\in(n,n+1)\\ 0, & x\notin(n,n+1)\end{cases}. \tag{4.8}$$

We have $\lim_{n\to\infty} f_n=0$ p.w., and yet for each n there exist values of x for which $f_n(x)=1$.

Uniform convergence, as the name implies, is defined to exclude the above possibilities and to ensure convergence uniformly at all points together instead of individually.

Definition A sequence $\{f_n\}$ is said to converge uniformly to f on a set S (consisting of the whole or some subset of the real line) if, for every $\varepsilon>0$, there exists an N, depending only on ε, such that whenever $n>N$

$$|f_n(x)-f(x)|<\varepsilon \qquad (\text{all } x\in S). \quad \square$$

Uniform convergence over a finite interval implies bounded convergence, so that (4.4) is valid. However, over an infinite interval uniform convergence does not imply dominated convergence, as the example of (4.8) shows: uniform convergence does not transfer integrability over an infinite range to the limit function.

From among the many consequences of uniform convergence, see for instance, Apostol (1974), Titchmarsh (1968), we pick out one, namely the transference of continuity.

Theorem 4.4 Consider a sequence $\{f_n\}$ which converges uniformly on S to f: if each f_n is continuous at some point x of S then f is continuous at x. □

Note that differentiability is not, however, transferred to the limit in the same way as continuity.

4.5 Convergence in the mean

We now break away from pointwise convergence and consider some quite different varieties of convergence which apply in an averaged or integrated fashion, and do not imply p.w. convergence.

Convergence in the mean is a crucial concept in treating the Fourier transforms of functions in L^p, as we shall see in later chapters. We define it thus.

Definition Consider a sequence $\{f_n\}_{n=1}^{\infty}$ and a function f, all these functions being L^p for some fixed p satisfying $1 \leqslant p < \infty$: we say the sequence converges to f as a limit in the mean with index p and write

$$\underset{n\to\infty}{\text{l.i.m.}}\,(p) f_n = f$$

if

$$\lim_{n\to\infty} \int |f_n(x) - f(x)|^p \, dx = 0. \quad \square \tag{4.9}$$

An alternative phraseology, when $\{f_n\}$ converges to f as a limit in the mean with index (or order) p, is to say that $\{f_n\}$ converges to f in L^p; the sequence itself is said to converge in the mean or to converge in L^p. In terms of the p-norm, (4.9) means that:

$$\lim_{n\to\infty} \| f_n - f \|_p = 0.$$

The following example shows that mean convergence does not imply pointwise convergence a.e. Consider a sequence $\{f_n\}$ in which each f_n has a graph consisting of a rectangular pulse of unit height centred somewhere in the interval $[0, 1]$, and suppose that as $n \to \infty$ so the pulse gets narrower and moves to and fro across the interval. One way of achieving this is to construct for $n = 1, 2, 3, \ldots,$

$$f_n(x) = \begin{cases} 1, & \dfrac{n-2^m}{2^m} \leqslant x \leqslant \dfrac{n+1-2^m}{2^m} \\ 0, & \text{otherwise} \end{cases} \tag{4.10}$$

where m is an integer defined for each n by

$$2^m \leqslant n < 2^{m+1}.$$

We find $\{f_n\}$ converges to zero in L^1, and yet at no point in (0, 1) does $f_n(x)$ tend to zero.

The limit function involved in mean convergence is uniquely determined apart from its value on a null set. This finds expression in the two following results.

Theorem 4.5 Consider a sequence of functions $\{f_n\}$ and two functions f and g, and suppose that for some $p \in [1, \infty)$

$$\underset{n\to\infty}{\text{l.i.m.}}(p) f_n = f$$

and

$$\lim_{n\to\infty} f_n = g \quad \text{p.w.a.e.:}$$

then it follows that $f = g$ a.e. $\square$

Theorem 4.6 Consider a sequence of functions $\{f_n\}$ and two functions f and g, and suppose that for some $p \in [1, \infty)$ and some $r \in [1, \infty)$

$$\underset{n \to \infty}{\text{l.i.m.}}(p) f_n = f$$

and

$$\underset{n \to \infty}{\text{l.i.m.}}(r) f_n = g:$$

then it follows that $f = g$ a.e. □

4.6 Cauchy sequences

There exists an important test which determines whether or not a sequence of functions $\{f_n\}$ will converge in the mean as $n \to \infty$. The test relies on the concept of a Cauchy sequence defined as follows.

Definition A sequence of functions $\{f_n\}_{n=1}^{\infty}$, each f_n belonging to L^p for some fixed p satisfying $1 \leqslant p < \infty$, is said to be a Cauchy sequence in L^p if for every $\varepsilon > 0$ there exists a positive integer N, depending only on ε, such that

$$\|f_n - f_m\|_p < \varepsilon$$

whenever $n > N$ and $m > N$. □

The test now consists in the following.

Theorem 4.7 Let $\{f_n\}_{n=1}^{\infty}$ be a sequence of functions in L^p for some p satisfying $1 \leqslant p < \infty$: then there will exist some limit function f to which the sequence converges, as a limit in the mean of index p, if and only if the sequence is a Cauchy sequence in L^p. □

One consequence of this theorem is that each Cauchy sequence in L^p converges in mean to a limit function that is also in L^p; because of this the space L^p, for each fixed $p \in [1, \infty)$, is said to be *complete*. This lies at the root of a rather more general approach to convergence in terms of metric spaces, see for instance Pitts (1972), Sutherland (1975).

5

Local averages and convolution kernels

5.1 Introduction

The idea behind various Fourier inversion theorems is that Fourier transformation of a function f to give F followed by inverse transformation of F will yield back the original function f. However, it can happen that at some point $x=x_0$ the inversion process will not yield the value $f(x_0)$ but rather a value based on some average of $f(x)$ around the point x_0. At a simple step discontinuity for instance a value 'half way up the step' may arise, independent of the value originally assigned to f at that point. Various different forms of the inversion theorem depend on local averages calculated in different ways, and the present chapter is devoted to a discussion of several forms of local average.

The most natural way of defining the locally averaged value of a locally integrable function f at the point x is through the limit

$$\lim_{\varepsilon\to 0}\frac{1}{2\varepsilon}\int_{x-\varepsilon}^{x+\varepsilon} f(u)du.$$

It follows from theorem 3.9, due to Lebesgue, that if f is locally integrable then the above limit will in fact exist for a.a.x, and will equal $f(x)$ a.e. We can use the nomenclature of convolution to express this another way. If r_n is the 'rectangle' function defined, for $n=1, 2, 3, \ldots$, by

$$r_n(x)=\begin{cases} n/2, & |x|<1/n \\ 0, & |x|\geqslant 1/n \end{cases}$$

and, if for each n, we form the convolution $f*r_n$, then

$$\lim_{n\to\infty} f*r_n=f \quad \text{p.w.a.e.}$$

We may consider using some other sequence $\{k_n\}_{n=1}^{\infty}$ of functions in place of the rectangle functions, and the following question naturally arises. Given a function f and a sequence $\{k_n\}_{n=1}^{\infty}$, such that the convolution $f*k_n$

is defined for all n, what conditions must f and the k_n jointly satisfy in order to ensure convergence (of some specified type such as a pointwise or as a limit in the mean) of $f * k_n$ to f as $n \to \infty$? A sequence $\{k_n\}$ of functions satisfying some such set of conditions is conveniently called a *convolution kernel*. Very often such a sequence is generated from a single function k by defining $k_n(x) = nk(nx)$, and in this case we refer to k as the convolution kernel, on the understanding that the sequence is formed in this way. Throughout, the discrete index n may be replaced by a continuous, real parameter λ and the limit $\lambda \to \infty$ or $\lambda \to \lambda_0$ used.

Historically the convolution kernel $k(x) = (\pi x)^{-1} \sin x$ ($=0$ when $x=0$) was very important in the early development of Fourier theory, and in honour of the associated work by P. G. L. Dirichlet (1805–59) this is often called the Dirichlet kernel for Fourier transforms. As we shall see the Dirichlet kernel has some awkward characteristics, and several other types of kernel have more convenient properties.

In the treatment of convolution kernels theorem 3.9 plays an important role, and we start with a further description of the consequences of this particular Lebesgue theorem.

5.2 Lebesgue points

It follows from theorem 3.9 that if $f \in L(a, b)$ then for almost all points x on (a, b) there will exist a complex number, $f_L(x)$ say, such that

$$\lim_{\varepsilon \to 0^+} \frac{1}{\varepsilon} \int_0^\varepsilon |f(x+u) + f(x-u) - 2f_L(x)| du = 0, \tag{5.1}$$

and that for a.a. such points $f_L(x) = f(x)$. It is convenient to call a point at which $f_L(x)$ exists a *Lebesgue point* of f, and to refer to $f_L(x)$ as the *Lebesgue value* of the function at that point. Note that the set of Lebesgue points defined in this way is larger than, and includes, the Lebesgue set as defined in section 3.5: the difference arises because $f(x)$ may be undefined or may differ from $f_L(x)$ at a Lebesgue point.

The Lebesgue value can correctly be thought of as a locally averaged value of the function, since if x is a Lebesgue point of f then

$$f_L(x) = \lim_{\varepsilon \to 0} \frac{1}{2\varepsilon} \int_{x-\varepsilon}^{x+\varepsilon} f(u) du. \tag{5.2}$$

However, the convergence of the limit on the right hand side of (5.2) is not in itself sufficient to ensure that x is a Lebesgue point of f; it is the modulus in the integrand of (5.1) that makes this a more stringent test of the local behaviour of f and leads to the importance of the Lebesgue points. As an example note that, although

$$\lim_{\varepsilon\to 0^+}\frac{1}{2\varepsilon}\int_{-\varepsilon}^{\varepsilon}\cos(1/x)\mathrm{d}x=0,$$

nevertheless the function $\cos(1/x)$, $x\neq 0$, does not have a Lebesgue point at $x=0$. The function $\sin(1/x)$, $x\neq 0$, does, however, have $f_{\mathrm{L}}(0)=0$, the existence of this Lebesgue point at $x=0$ arising from the odd symmetry of this function about $x=0$.

At a point of continuity of f it is true that $f_{\mathrm{L}}(x)=f(x)$. Also, at any point where $f(x^+)$ and $f(x^-)$ are defined, (2.3), then

$$f_{\mathrm{L}}(x)=\tfrac{1}{2}\{f(x^+)+f(x^-)\}.$$

However, $f(x^+)$ and $f(x^-)$ are not necessarily defined at a Lebesgue point; for instance, if $f(x)=|x|^{-1/2}\,\mathrm{sgn}\,x$ ($=0$ when $x=0$) then $f_{\mathrm{L}}(0)=0$ despite the infinite discontinuity at $x=0$. Here and subsequently we define the signum function by

$$\mathrm{sgn}\,x=\begin{cases}1, & x>0\\ 0, & x=0\\ -1, & x<0.\end{cases}\tag{5.3}$$

The same ideas can be applied to the behaviour of a function just to the right or left of a point. Thus if $f\in L(a,b)$ then for almost all points x within (a,b) there will exist a complex number, say $f_{\mathrm{L}+}(x)$, such that

$$\lim_{\varepsilon\to 0^+}\int_0^{\varepsilon}|f(x+u)-f_{\mathrm{L}+}(x)|\mathrm{d}u=0,$$

and at almost all such points $f_{\mathrm{L}+}(x)=f(x)$; likewise at almost all points within (a,b) there will exist a complex number, say $f_{\mathrm{L}-}(x)$, such that

$$\lim_{\varepsilon\to 0^+}\int_0^{\varepsilon}|f(x-u)-f_{\mathrm{L}-}(x)|\mathrm{d}u=0,$$

and at almost all such points $f_{\mathrm{L}-}(x)=f(x)$. It is convenient to refer to $f_{\mathrm{L}+}(x)$ and $f_{\mathrm{L}-}(x)$ as the right and left hand Lebesgue values of f at the point x.

As might be expected the existence of $f(x^+)$ implies the existence of $f_{\mathrm{L}+}(x)$, in which case they have the same value, but $f_{\mathrm{L}+}(x)$ may exist when $f(x^+)$ is undefined; similarly with left hand values. Further if $f_{\mathrm{L}+}(x)$ and $f_{\mathrm{L}-}(x)$ exist at some point, then so also will $f_{\mathrm{L}}(x)$ and

$$f_{\mathrm{L}}(x)=\tfrac{1}{2}\{f_{\mathrm{L}-}(x)+f_{\mathrm{L}+}(x)\}.$$

However, the existence of $f_{\mathrm{L}}(x)$ does not imply the existence of $f_{\mathrm{L}-}(x)$ or $f_{\mathrm{L}+}(x)$, as the function $|x|^{-1/2}\,\mathrm{sgn}\,x$ shows. Consider also the function whose graph consists of rectangular pulses centred on $x=1/n$, $n=1,2,3,\ldots$, each pulse having unit height and a width $1/n^3$. We find

$$f_L(0)=f_{L-}(0)=f_{L+}(0)=0, \tag{5.4}$$

whilst $f(0^+)$ and $f(0^-)$ are undefined.

5.3 Approximate convolution identities

An important class of convolution kernel is defined as follows.

Definition Let $\{k_\lambda\}_{\lambda=1}^{\infty}$ be a sequence of real or complex valued functions, with either a discrete or continuous parameter λ: we call the sequence an *approximate convolution identity* if it satisfies the following three conditions:

(i) $k_\lambda \in L$ and $\lim_{\lambda\to\infty} \int k_\lambda = 1$,

(ii) there exists a single $A>0$ such that for each λ

$$\int |k_\lambda| < A,$$

(iii) for any $\delta>0$,

$$\lim_{\lambda\to\infty} \int_{|x|>\delta} |k_\lambda(x)|\mathrm{d}x=0. \quad \square$$

Such sequences are easy to construct since given any $g\in L$, the sequence $\{g_\lambda\}_{\lambda=1}^{\infty}$, where $g_\lambda(x)=\lambda g(\lambda x)$ a.e., will be an approximate convolution identity. If g is real and has a graph consisting of a hump centred on the origin, then the graph of g_λ consists of a hump which becomes taller and narrower as $\lambda\to\infty$, the area remaining constant. In the language of generalized functions such a sequence provides one way of representing a delta function. We give some examples at the end of this section.

The importance of approximate convolution identities stems from the following two theorems, relating to mean convergence and pointwise convergence, respectively.

Theorem 5.1 Suppose $f\in L^p$ for some p satisfying $1\leqslant p<\infty$, and let $\{k_\lambda\}_{\lambda=1}^{\infty}$ be an approximate convolution identity: then the convolution $f*k_\lambda$ is defined a.e. for each λ and is in class L^p, and

$$\underset{\lambda\to\infty}{\mathrm{l.i.m.}}(p) f*k_\lambda = f. \quad \square$$

The second theorem is complementary to this, and covers the case $p=\infty$.

Theorem 5.2 Suppose $f\in L^\infty$, and let $\{k_\lambda\}_{\lambda=1}^{\infty}$ be an approximate convolution identity: then for each λ the convolution $f_\lambda=f*k_\lambda$ is everywhere continuous, and (i) at each point of continuity of $f(x)$

$$\lim_{\lambda\to\infty} f_\lambda(x)=f(x),$$

(ii) if $f(x)$ is continuous on a finite interval (a, b) then on any interior interval $[c, d]$, $a<c<d<b$,

$$\lim_{\lambda\to\infty} f_\lambda(x)=f(x), \qquad \text{uniformly on } [c, d],$$

and (iii) if $f(x)$ is continuous on $(-\infty, \infty)$ and tends to zero as $x\to\pm\infty$, then f_λ tends uniformly to f on $(-\infty, \infty)$ as $\lambda\to\infty$. □

If the conditions on the convolution kernel are tightened, then it is possible to relax the conditions on f and also to guarantee convergence at simple step discontinuities.

***Theorem* 5.3** Suppose the function f is such that $f(x)/(1+|x|)$ is integrable over $(-\infty, \infty)$, and let k be a real or complex valued function such that (*a*) $k\in L$ and $\int k=1$, (*b*) k is bounded on $(-\infty, \infty)$, (*c*) $xk(x)\to 0$ as $x\to\pm\infty$: then (i) the convolution

$$f_\lambda(x)=f*k_\lambda(x) \tag{5.5a}$$

is defined for all $\lambda>0$ and all real x, where

$$k_\lambda(x)=\lambda k(\lambda x), \tag{5.5b}$$

(ii) at each point x at which $f(x^+)$ and $f(x^-)$ are defined,

$$\lim_{\lambda\to\infty} f_\lambda(x)=f(x^-)\int_0^\infty k(u)du+f(x^+)\int_{-\infty}^0 k(u)du,$$

(iii) if in addition $k(x)=k(-x)$ a.e. then

$$\lim_{\lambda\to\infty} f_\lambda(x)=\lim_{\varepsilon\to 0^+}[f(x+\varepsilon)+f(x-\varepsilon)]$$

at each point at which the right hand limit exists. □

By constraining the kernel yet further it is possible to achieve convergence almost everywhere, as in the following.

***Theorem* 5.4** Let $k(x)$ be a function defined at all x such that (*a*) $k\in L$ and $\int k=1$, (*b*) there exists an $A>0$ and an $s>1$ such that for all x

$$|k(x)|<A(1+|x|)^{-s};$$

suppose also that f is a function such that the convolution

$$f_\lambda(x)=f*k_\lambda(x)$$

is defined for each $\lambda>0$ and each real x, where k_λ is defined at all x by

$$k_\lambda(x)=\lambda k(\lambda x):$$

then it will follow that: (i) on the Lebesgue set of f, and so at a.a.x,

$$\lim_{\lambda\to\infty} f_\lambda(x)=f(x)$$

(ii) at each point where $f_{L-}(x)$ and $f_{L+}(x)$ are defined, and so at a.a.x,

$$\lim_{\lambda\to\infty} f_\lambda(x)=f_{L-}(x)\int_0^\infty k(u)\mathrm{d}u+f_{L+}(x)\int_{-\infty}^0 k(u)\mathrm{d}u$$

(iii) if in addition $k(x)=k(-x)$ at all x, then

$$\lim_{\lambda\to\infty} f_\lambda(x)=f_L(x)$$

at each point where $f_L(x)$ is defined, and so a.e. □

The convolution kernel $k(x)=[\pi(1+x^2)]^{-1}$ occurs frequently in Fourier theory and is associated with the work of Poisson, whilst the kernel $k(x)=[(\sin \pi x)/\pi x]^2$ is known as the Fejér convolution kernel: both these kernels satisfy the conditions in theorems 5.3 and 5.4. Other kernels which also satisfy theorems 5.3 and 5.4 are the functions $\frac{1}{2}e^{-|x|}$, $\exp(-\pi x^2)$, $\frac{1}{2}r(x)$, $(1-|x|)r(x)$, where r is the rectangle function defined by

$$r(x)=\begin{cases}1, & -1<x<1\\ 0, & |x|\geqslant 1\end{cases}. \tag{5.6}$$

5.4 The Dirichlet kernel and Dirichlet points

The Dirichlet kernel for Fourier transforms, $(\sin x)/\pi x$, decays too slowly at infinity to satisfy the conditions in any of the theorems 5.2–5.4 just described. However, if $f\in L^p$ for some p satisfying $1\leqslant p<\infty$, or more generally if

$$\int_{-\infty}^\infty \frac{f(x)}{1+x}\mathrm{d}x<\infty,$$

then the convolution

$$f_\lambda(x)=\int_{-\infty}^\infty f(x-u)\frac{\sin \lambda u}{\pi u}\mathrm{d}u \tag{5.7}$$

is defined for all real x at each $\lambda>0$. The question then arises, in what way, if at all, will f_λ tend to f as $\lambda\to\infty$?

Mean convergence is guaranteed in some cases, as follows.

Theorem 5.5 Suppose $f\in L^p$ for some p satisfying $1<p<\infty$ and let f_λ be defined by (5.7): then it follows that

$$\underset{\lambda\to\infty}{\text{l.i.m.}(p)} f_\lambda=f. \quad \square$$

This theorem fails for $p=1$ and $p=\infty$, though theorem 5.11 will describe some limited results in this direction.

Pointwise convergence of $f_\lambda(x)$ to $f(x)$ at a particular value of x, as $\lambda\to\infty$, depends on the behaviour of f in an arbitrarily small interval surrounding the point x, and it is convenient to make the following definitions.

Definition Consider a function $f(x)$ and suppose that for some $\delta > 0$ and some fixed value of x,

$$\lim_{\lambda\to\infty} \int_{x-\delta}^{x+\delta} f(x+u)\,\frac{\sin \lambda u}{\pi u}\,du \tag{5.8}$$

converges to a limiting value, say $f_D(x)$: then we will say that x is a *Dirichlet point* of $f(x)$, and we call $f_D(x)$ the *Dirichlet value* of f at the point x. □

Note that if (5.8) converges for one value of δ, then it will converge also, to the same limit, for all lesser positive values of δ, so that δ may be chosen arbitrarily small. If f is continuous at a Dirichlet point x, then $f_D(x) = f(x)$; also, if $f(x^+)$ and $f(x^-)$ are defined at a Dirichlet point, then

$$f_D(x) = \tfrac{1}{2}[f(x^+) + f(x^-)].$$

Since $f_D(x)$ represents the value of a form of local average of f around the point x it may be that $f_D(x)$ is defined at a point where $f(x)$ is itself undefined.

Pointwise convergence of the Dirichlet convolution now depends on the following.

Theorem 5.6 Suppose f is locally integrable and that

$$\int_{-\infty}^{\infty} \frac{f(x)}{1+|x|}\,dx < \infty,$$

and let the convolution $f_\lambda(x)$ be defined by (5.7): then

$$\lim_{\lambda\to\infty} f_\lambda(x)$$

will converge if, and only if, x is a Dirichlet point of f, in which case the value of the limit will be equal to $f_D(x)$. □

It is a surprising fact that continuity of a function at a point is neither sufficient nor necessary to ensure that the point is a Dirichlet point, though continuity of f at all points on an interval does ensure that almost all points on the interval are Dirichlet points. It is likewise a disconcerting fact that there are integrable functions which possess no Dirichlet points. These matters are elaborated in sections 5.5–5.7. One naturally seeks simple conditions that are sufficient to establish a Dirichlet point, and in sections 5.8–5.11 we describe various such conditions.

5.5 The functions of du Bois-Reymond and of Fejér

It was first shown by du Bois-Reymond in 1872 that a function could be constructed that was continuous everywhere on $(-\pi, \pi)$ but which failed to have a Dirichlet point at $x=0$. Fejér gave further examples and it is now known that a continuous function may fail to have a Dirichlet value at a denumerable or even a non-denumerable infinity of points in every finite interval.

These functions are described for instance by Edwards (1979) and we will quote here one of the simplest examples, due to Fejér. Let the variables m and n represent positive integers, let x be any real number and construct functions k, g and f as follows:

$$\left.\begin{aligned} k(n) &= 2^{n^3} \\ g(n,x) &= \sum_{m=1}^{k(n)} \frac{\sin mx}{m} \\ f(x) &= \sum_{n=1}^{\infty} \{n^{-2}\sin[2xk(n)]g(n,x)\}. \end{aligned}\right\} \tag{5.9}$$

The series defining f converges at all real x, and f is everywhere continuous, is periodic with period 2π, and has $f(0)=f(0^-)=f(0^+)=0$: however, the point $x=0$ is *not* a Dirichlet point of $f(x)$. This behaviour is associated with the nature of oscillations in the graph of $f(x)$ as $x\to 0$.

5.6 Carleson's theorem

Whether or not a continuous function has Dirichlet points almost everywhere remained one of the outstanding problems associated with Fourier theory until, in 1966, an affirmative answer was given by Carleson (1966). Carleson showed, more generally, that a locally square integrable function has Dirichlet points a.e., and Hunt (1968) then showed that a function that is pth power integrable locally will have Dirichlet points a.e. when $p>1$. We combine these results as follows.

***Theorem* 5.7** Consider a function f that is $L^p(a,b)$ for some p satisfying $1<p\leqslant\infty$ for some finite or infinite interval (a,b): then a.e. point x on (a,b) will be a Dirichlet point of $f(x)$, and $f(x)=f_D(x)$ a.e. □

This result means not only that continuous functions have Dirichlet points a.e., but also that any function that is integrable in the Riemann sense (as the sum of a finite number of improper Riemann integrals) will possess Dirichlet points a.e.

The omission of the case $p=1$ from the Carleson–Hunt theorem is significant, as we see in the next section.

5.7 Kolmogoroff's theorem

In 1926 A. N. Kolmogoroff constructed a periodic, locally integrable function that possessed *no* Dirichlet points: on multiplying this function by, for example, the function $\exp(-x^2)$ we obtain a function that is $L(-\infty, \infty)$ and possesses no Dirichlet points. We thus have the following.

***Theorem* 5.8** Given any finite or infinite interval (a, b) there will exist functions that are $L(a, b)$ and which possesses no Dirichlet points on (a, b). □

The construction of the Kolmogoroff function is involved and is described for instance on pp. 310–14 of Zygmund (1959). Such a function defies any attempt to draw its graph, since it is neither continuous nor bounded, on any finite interval, even though a well defined value exists almost everywhere. The Kolmogoroff function is not Riemann integrable over any finite interval, nor is it L^p over any interval for any $p>1$.

Whilst in a sense the examples of du Bois-Reymond, Fejér and Kolmogoroff emphasize the failings of the Dirichlet kernel, the next theorems, of Jordan, Dini and of de la Vallée-Poussin work in the opposite direction and give sufficient (but not necessary) conditions that a point will be a Dirichlet point of a function. Rather than quote the most powerful of the results first (that of de la Vallée-Poussin) we prefer to work towards it through a set of simpler conditions.

5.8 The Dirichlet conditions

Dirichlet discovered a simple but very useful condition on a function f which is sufficient to ensure that a particular point is a Dirichlet point of f. The condition is based on the idea of a piecewise monotonic function, and as a preliminary we define this.

A function f that is real valued on some set I of points on the real line is said to be *montonically increasing* (or non-decreasing) on I if, for points x_1 and x_2 on I, $x_1 < x_2$ implies $f(x_1) \leqslant f(x_2)$; the function is said to be *strictly* monotonically increasing on I if $x_1 < x_2$ implies $f(x_1) < f(x_2)$. Likewise the terms *monotonically decreasing* (or non-increasing) and *strictly* monotonically decreasing are defined when the inequality $x_1 < x_2$ in the above definitions is replaced by $x_1 > x_2$. A function is said to be *monotonic* on I if it is either monotonically increasing or decreasing on I. A function f is said to be *piecewise monotonic* on an interval $[a, b]$ if f is defined and real valued at each point on $[a, b]$, and if $[a, b]$ can be partitioned into a finite number N of open sub-intervals, (a, a_1), $(a_1, a_2), \ldots, (a_N, b)$, where $a < a_1 < a_2 < \cdots < a_N < b$, such that f is monotonic on each sub-interval.

Dirichlet's condition runs as follows.

Theorem 5.9 Let $f(x)$ be a real or complex valued function satisfying the following conditions on some finite interval $[a, b]$, (i) f is defined at each point of $[a, b]$, (ii) f is bounded on $[a, b]$, (iii) f has at most a finite number of discontinuities on $[a, b]$, and (iv) the real and imaginary parts of f are each piecewise monotonic on $[a, b]$: then it follows that each point of (a, b) is a Dirichlet point of f, and that for all points on (a, b) the limits $f(x^+)$ and $f(x^-)$ exist and

$$f_D(x)=\tfrac{1}{2}[f(x^+)+f(x^-)]. \quad \square$$

We will say that a function satisfying conditions (i)–(iv) in the above theorem satisfies the *Dirichlet conditions* on $[a, b]$.

5.9 Jordan's theorem

The Dirichlet conditions are a special case of a less stringent condition due to C. Jordan (1838–1922). To describe Jordan's condition we need to introduce the concept of bounded variation.

When we say that a real valued function is of bounded variation on some neighbourhood of a point $x=x_0$ (or, more colloquially, around x_0) we mean roughly speaking that the graph of f does not oscillate up and down through an infinite distance as x is scanned through the value x_0. Thus if f is of bounded variation (b.v.) around x_0, then $f(x)$ cannot tend to infinity as $x \to x_0$; however, a finite step discontinuity at $x=x_0$ is allowed. Likewise if a function oscillates infinitely rapidly at constant amplitude around a point, as does $\cos(1/x)$ around $x=0$, then it will not be of b.v. around the point. However, the function $x^2\cos(1/x)$, ($=0$ at $x=0$), is of b.v. around $x=0$ because the amplitude of the oscillations tends to zero fast enough as $x \to 0$. A function may be both continuous and differentiable at a point, and yet not be of b.v. around the point, as is the case at $x=0$ with the function

$$f(x)=\begin{cases} x^2\cos(1/x^2), & x\neq 0 \\ 0, & x=0 \end{cases}. \tag{5.10}$$

A strict definition of bounded variation runs as follows.

Definition A function f with real or complex values defined at all points on the finite closed interval $[a, b]$ is said to be of bounded variation on $[a, b]$ if a single number $A>0$ exists such that

$$\sum_{n=1}^{N} |f(x_n)-f(x_{n-1})| < A \tag{5.11}$$

for every finite set of numbers $x_0, x_1, x_2, \ldots, x_N$ satisfying

$$a \leqslant x_0 < x_1 < x_2 < \cdots < x_N \leqslant b;$$

we say that a function f is of bounded variation on some neighbourhood of a point $x=x_0$ if f is of bounded variation on $[x_0-\delta, x_0+\delta]$ for some $\delta>0$; we say f is of bounded variation on $(-\infty, \infty)$ if (5,11) holds, with a single $A>0$, for every finite set of numbers $\{x_n\}_{n=0}^{N}$ satisfying

$$x_0<x_1<x_2<\cdots<x_N. \quad \square$$

The idea behind Jordan's theorem is that if a function f is of b.v. around some point x, then $f(x^+)$ and $f(x^-)$ will both be defined at x, also x will be a Dirichlet point of f and

$$f_D(x)=\tfrac{1}{2}[f(x^-)+f(x^+)].$$

However, one may exploit the fact that the Dirichlet kernel, $\sin x/\pi x$, is an even function to extend this result to cases when the symmetrical function

$$\varphi(u)=\tfrac{1}{2}[f(x_0+u)+f(x_0-u)] \tag{5.12}$$

is of b.v. around $u=0$, at fixed x_0, even when $f(x)$ is not itself of b.v. around $x=x_0$. We now state Jordan's result as follows.

Theorem 5.10 Suppose f is defined on $[x_0-\delta, x_0+\delta]$ and is $L(x_0-\delta, x_0+\delta)$ for some fixed x_0 and $\delta>0$, and suppose that $\varphi(u)$, defined on $u\in[-\delta, \delta]$ by (5.12), is of b.v. around $u=0$: then x_0 is a Dirichlet point of f and

$$f_D(x_0)=\lim_{u\to 0}\frac{f(x_0+u)+f(x_0-u)}{2}. \quad \square$$

For example, the function $|x|^{-1/2}\operatorname{sgn} x$ ($=0$ at $x=0$) satisfies the conditions of theorem 5.10 around $x=0$, and has $f_D(0)=0$.

The concept of b.v. has links with other properties of a function, the most fundamental link being that a real valued function f will be of b.v. on an interval $[a, b]$ if, and only if, f can be expressed on $[a, b]$ as the difference between two functions that are monotonic on $[a, b]$. It follows that if f is of b.v. on $[a, b]$ then f is integrable over $[a, b]$ as a proper Riemann integral, and that f is continuous a.e. on $[a, b]$. If a function f is absolutely continuous on $[a, b]$, or equivalently if it is an indefinite integral on $[a, b]$, then f will be of b.v. on $[a, b]$, so that every point on (a, b) will be a Dirichlet point of f.

The condition of b.v. also ensures more than simple pointwise convergence of convolution with the Dirichlet kernel, as the following shows.

Theorem 5.11 Suppose $f\in L_{\text{LOC}}$ is such that

$$\int_{-\infty}^{\infty}\frac{|f(x)|}{1+|x|}\,dx<\infty,$$

let $f_\lambda(x)$ be the convolution defined by (5.7), and suppose that f is of bounded variation on a finite interval $[a, b]$: then (i)

$$\lim_{\lambda\to\infty} \int_a^b |f_\lambda(x)-f(x)|\mathrm{d}x=0$$

and (ii) if additionally f is continuous on (a, b) then $f_\lambda(x)$ converges uniformly to $f(x)$ on any interval $[c, d]$ with $a<c<d<b$. □

5.10 Dini's theorem

Although continuity of f at a point x is not sufficient to ensure that x is a Dirichlet point of f, differentiability of f at x is a sufficient condition. This differentiability condition is a special case of a less stringent condition due to U. Dini (1845–1918). Just as differentiability at a point and b.v. around the point are independent conditions, neither implying the other, so also the Dini and Jordan conditions are independent.

In the following theorem we lead up to the Dini condition through a sequence of conditions, each more general than its predecessor. Equation (5.14) is known as a Lipschitz condition, and (5.15) is the Dini condition.

Theorem 5.12 Suppose $f\in L(x-\delta, x+\delta)$ for some fixed real x and some $\delta>0$, and define

$$\varphi(u)=\tfrac{1}{2}[f(x+u)+f(x-u)]$$

for all values of u on $[0, \delta]$ at which this expression is defined: if now the limit $\varphi(0^+)$ exists and the following form of derivative exists:

$$\lim_{u\to 0^+}\left[\frac{\varphi(u)-\varphi(0^+)}{u}\right],$$

then x is a Dirichlet point of f and

$$f_{\mathrm{D}}(x)=\lim_{u\to 0^+}\left[\frac{f(x+u)+f(x-u)}{2}\right]; \tag{5.13}$$

more generally, if $\varphi(0^+)$ exists and if real numbers α, β and B exist satisfying $\alpha>0$, $0<\beta<\delta$, $B>0$, such that for $u\in(0, \beta)$

$$|\varphi(u)-\varphi(0^+)|<Bu^\alpha, \tag{5.14}$$

then x is a Dirichlet point of f and (5.13) is true; finally, and more generally still, if a complex number A and an $\varepsilon>0$ exist such that

$$\int_0^\varepsilon \frac{|\varphi(u)-A|}{u}\,\mathrm{d}u \tag{5.15}$$

exists, then x is a Dirichlet point of f and

$$f_{\mathrm{D}}(x)=A=\lim_{\gamma\to 0^+}\frac{1}{2\gamma}\int_{-\gamma}^{\gamma} f(x+u)\mathrm{d}u. \quad \square$$

In contrasting the Jordan and Dini conditions we note that Dini's condition is based on an integral, so that rapid fluctuations in the value of f near x are less important than in Jordan's condition; on the other hand the u appearing in the denominator of the integrand in (5.15) can make the Dini condition more stringent than that of Jordan in some cases. For example, consider

$$f(x)=\begin{cases}0, & x=0\\ \dfrac{1}{\ln(1/|x|)}, & x\neq 0:\end{cases} \tag{5.16}$$

this function satisfies the Jordan condition around $x=0$ and

$$f_D(0)=0=f(0^+)=f(0^-);$$

however, the function converges to zero too slowly as $x\to 0$ for Dini's condition to be satisfied. On the other hand, consider

$$g(x)=\begin{cases}0, & x=0\\ |x|^{1/2}\cos(1/x), & x\neq 0:\end{cases} \tag{5.17}$$

this satisfies the Lipschitz and Dini conditions at $x=0$ but oscillates too strongly to satisfy Jordan's condition. A 'worse' function has a graph consisting of an infinite set of positive rectangular pulses centred at $x=\pm 2^{-n}$, $n=1,2,3,\ldots$, the height of a pulse being n and its width $(n^{-3}2^{-n})$, the function being elsewhere equal to zero. This function satisfies Dini's condition at $x=0$, and $f_D(0)=0$, but neither the Lipschitz nor the Jordan condition is satisfied.

The idea of a piecewise smooth function provides a simple condition for ensuring that every point on an interval satisfies Dini's condition. We define and use the concept as follows.

Theorem 5.13 Suppose f is continuous at all but a finite number of points on the closed interval $[a, b]$, and suppose also that (i) whenever $a\leqslant x<b$ then $f(x^+)$ is defined and the following right-hand derivative converges:

$$\lim_{\delta\to 0^+}\left[\frac{f(x+\delta)-f(x^+)}{\delta}\right],$$

and (ii) whenever $a<x\leqslant b$ then $f(x^-)$ exists and the following left-hand derivative converges:

$$\lim_{\delta\to 0^+}\left[\frac{f(x^-)-f(x-\delta)}{\delta}\right]:$$

then we say that f is *piecewise smooth* on $[a, b]$, and it will follow that every point on (a, b) is a Dirichlet point of f. □

5.11 The de la Vallée-Poussin test

An even more general condition for ensuring that a point is a Dirichlet point of a function is due to de la Vallée-Poussin, this condition being certainly satisfied when either the Dini or Jordan conditions are satisfied. The ideas of Jordan and Dini are brought together by applying a bounded variation test to an averaged version of the function rather than to the function itself.

Theorem 5.13 Suppose a function f is such that, for some fixed x and some $\delta > 0$, the following conditions are satisfied: (i) $f \in L(x-\delta, x+\delta)$, and (ii) the function $\psi(u)$, defined on $[0, \delta]$ by

$$\psi(u) = \begin{cases} \dfrac{1}{2u} \displaystyle\int_{-u}^{u} f(x+t)\mathrm{d}t, & 0 < u \leqslant \delta \\ 0, & u = 0 \end{cases}$$

is of b.v. on $[0, \delta]$: then x will be a Dirichlet point of f and

$f_{\mathrm{D}}(x) = \psi(0^+)$. □

This concludes our summary of the commonly quoted tests for the existence of a Dirichlet point: no simple test that is both necessary and sufficient is known.

6

Some general remarks on Fourier transformation

6.1 Introduction

In this chapter we make some general remarks about Fourier transformation before proceeding to a systematic statement of theorems in the remaining chapters. The aim will be to progress from simple situations towards more complicated ones, and we start in chapter 7 with the so-called good functions. We then progressively consider wider classes of functions and at each stage show in what way the theorems need to be modified. Throughout chapters 7–11 we retain the restriction that both the function and its Fourier transform are locally integrable functions defined almost everywhere on the real line; such pairs we will call *ordinary Fourier pairs*. Throughout we will assume the use of Lebesgue integration (see chapter 2) unless otherwise stated, and the functions in an ordinary Fourier pair can display fairly 'pathological' local behaviour. However, it is possible to consider 'even worse' situations, and in chapters 12–14 we show how Fourier transformation can be applied to entities which are not locally integrable and are not even defined through their values almost everywhere on the real line. These entities are the generalized functions, and we will introduce several examples of these including the well known Dirac delta function.

Although in stating theorems it is necessary to distinguish clearly between those which apply to generalized functions and those which apply to ordinary functions, we wish to stress the view that the introduction of generalized functions is only one step in a sequence of generalizations, and that the ordinary pairs can be regarded as special cases within the generalized theory.

6.2 The definition of the Fourier transform

The definition of what is meant by the Fourier transform of a given function is not so straightforward as might be expected. In fact we will progressively modify and generalize the definition as different classes of function or generalized function are considered. Consider for example the first step in this process. If a function f is $L(-\infty, \infty)$ then the following integral will exist for all y and may be used to define a function F thus,

$$F(y) = \int_{-\infty}^{\infty} f(x) \exp(-2\pi i x y) \mathrm{d}x; \tag{6.1}$$

we call F the Fourier transform of f, and we call f the inverse Fourier transform of F, whilst the term $\exp(-2\pi i x y)$ is called the Fourier kernel. However, the integral in (6.1) will not converge, as a Lebesgue integral, for any value of y if f is not $L(-\infty, \infty)$ and to proceed beyond the class L it will be necessary to modify the definition: this would apply for instance to the function defined a.e. as $\sin x/x$, which is $L^2(-\infty, \infty)$ but not $L(-\infty, \infty)$. One possibility, amongst several, is a definition based on the following limit, which for $f \in L^2(-\infty, \infty)$ will converge for almost all y,

$$F(y) = \lim_{\lambda \to \infty} \int_{-\infty}^{+\infty} \exp(-|x|/\lambda) f(x) \exp(-2\pi i x y) \mathrm{d}x.$$

Each successive definition will be consistent with its predecessors, will be applicable to a wider class of functions, and each will be connected in some way with the integral in (6.1); the connection will become less direct as we progress. The notation $f \leftrightarrow F$ will be used to indicate that the entity to the right of the arrow is the Fourier transform of that on the left, and that the entity to the left of the arrow is the inverse transform of that on the right.

The operation of Fourier transformation is not strictly unique, since if the function f in (6.1) is replaced by another differing in value from f on a null set but nevertheless equal a.e. to f, then the same function F will emerge as the transform; two different functions thus have the same transform. However, it is always true with ordinary Fourier pairs that if we class together functions which are equal a.e. to each other, then Fourier transformation is unique in that one such class will be associated with just one other such class as its transform or inverse transform. Thus when we say, for example, that F, defined for all y by $F(y) = [\pi(1+y^2)]^{-1}$ is the Fourier transform of f, defined for all x by $f(x) = \exp(-2\pi|x|)$, we mean that F together with all those functions equal a.e. to F are jointly the Fourier transform of those functions equal a.e. to f. This use of language is convenient, and is consistent with a form of uniqueness applicable to generalized functions.

There are other matters of notation. A real or complex valued function defined a.e. on the real line, or over part or all of the complex plane, will be called an *ordinary function*. If a letter, such as f, is used to represent an ordinary function, then the real or complex number that f assigns to the number x within its domain of definition will be written $f(x)$ and is called the function value of f at x. The variables x and y will be used throughout to represent real variables spanning the domain $(-\infty, \infty)$. Since an ordinary Fourier pair represents a relation between functions rather than function values it is more appropriate to write $f \leftrightarrow F$ than $f(x) \leftrightarrow F(y)$ to indicate such a pair. Nevertheless it is often convenient to represent an ordinary function by a mathematical expression giving its function values at a.a.x (or a.a.y) on $(-\infty, \infty)$, and with this understanding the following, for example, has a clear meaning,

$$\exp(-2\pi|x|) \leftrightarrow [\pi(1+y^2)]^{-1}.$$

As another example, if $g \leftrightarrow G$, where g and G are defined in terms of ordinary functions f and F as follows, for some fixed real a,

$$g(x) = f(x-a), \qquad \text{(a.a.}x\text{)}$$

$$G(y) = \exp(-2\pi i a y)F(y), \qquad \text{(a.a.}y\text{)}$$

then we will, for convenience, write this as,

$$f(x-a) \leftrightarrow \exp(-2\pi i a y)F(y).$$

Certain theorems on linearity, shifting, scaling, complex conjugation and transposition apply for all ordinary Fourier pairs. We will therefore state them now, once and for all. Complex conjugation is represented by an asterisk, whilst a subscript T will indicate transposition, so that the ordinary function f_T is defined from the ordinary function f by the requirement that

$$f_T(x) = f(-x) \tag{6.2}$$

at each value of x for which $f(-x)$ is defined.

Theorem 6.1 Suppose that $f \leftrightarrow F$ and $g \leftrightarrow G$ are ordinary Fourier pairs: then for each pair of complex numbers α and β, for every real number a, and for every non-zero real number b it follows that,

linearity theorem

$$(\alpha f + \beta g) \leftrightarrow (\alpha F + \beta G), \tag{6.3}$$

shift theorem

$$f(x-a) \leftrightarrow \exp(-2\pi i a y)F(y) \tag{6.4}$$

and

$$\exp(2\pi i a x)f(x) \leftrightarrow F(y-a), \tag{6.5}$$

scaling theorem

$$\left.\begin{array}{l} |b|f(bx) \leftrightarrow F(y/b) \\ f(x/b) \leftrightarrow |b|F(by), \end{array}\right\} \tag{6.6}$$

complex conjugation and transposition

$$\left.\begin{array}{ll} f^* \leftrightarrow F_T^*, & f_T \leftrightarrow F_T, \\ f_T^* \leftrightarrow F^*, & F \leftrightarrow f_T. \end{array}\right\} \quad \square \tag{6.7}$$

Each of the formulae (6.2)–(6.7) remains valid when f and F are replaced by generalized functions.

Various alternative conventions exist for defining the Fourier transform, and these differ from (6.1) in having a different factor in the exponent of the Fourier kernel, and/or a different factor in front of the integral. All these cases are covered in the following, where A and b are non-zero real numbers,

$$F(y) = A \int_{-\infty}^{\infty} f(x) \exp(-ibxy) \mathrm{d}x. \tag{6.8}$$

The convention in this book corresponds to $A = 1$, $b = 2\pi$, though the choice $A = 1$, $b = -2\pi$ is often met. In the physical sciences the cases $A = 1$, $b = \pm 1$ are common, whilst in mathematical contexts the cases $A = (2\pi)^{-1/2}$, $b = \pm 1$ are common. The inversion and convolution theorems in the general case lead to the following formal expressions in place of (1.1) and (1.2),

$$f(x) = \frac{|b|}{2\pi A} \int_{-\infty}^{\infty} F(y) \exp(ibxy) \mathrm{d}y \tag{6.9}$$

$$\int_{-\infty}^{\infty} f(x')g(x - x')\mathrm{d}x' = \frac{|b|}{2\pi A^2} \int_{-\infty}^{\infty} F(y)G(y) \exp(ibxy) \mathrm{d}y. \tag{6.10}$$

The choices $A = 1$, $b = \pm 2\pi$, are the only ones that lead to no factor before the integral sign in each of (6.8), (6.9) and (6.10). So far as matters of convergence are concerned it is of course quite immaterial which convention is adopted.

6.3 Sufficient conditions for transformability

So far as the scope of this book is concerned, for an entity to possess a Fourier transform it is sufficient and necessary that the entity should be a generalized function. Since the class of generalized functions will be constructed to include all locally integrable functions, this means that any locally integrable function will have a transform assigned to it, though the transform will not necessarily be an ordinary function and may even be an ultradistribution (a type of generalized function not always described in introductory treatments – see chapter 14).

If we restrict consideration to ordinary Fourier pairs this simplicity is lost, and it is a notorious feature of the subject that no direct and simple criterion is known which is both sufficient and necessary in ensuring that a function belongs to an ordinary Fourier pair. Certain very useful sufficient conditions can be given however, as well as some necessary conditions and we describe the simplest of these now.

Probably the most widely used sufficient condition is the following. If a function belongs to $L^p(-\infty,\infty)$ for some p satisfying $1 \leqslant p \leqslant 2$, then it will possess an ordinary Fourier transform and some form of inversion theorem will be applicable. The importance of this result lies in the fact that so many of the functions appearing in scientific applications are either integrable, $p=1$, or square integrable, $p=2$. As examples consider the ordinary function $|x|^{-1/2}\exp(-|x|)$ which is $L(-\infty,\infty)$, but not $L^2(-\infty,\infty)$, and the ordinary function $\sin x/x$ which is $L^2(-\infty,\infty)$, but not $L(-\infty,\infty)$. Both belong to ordinary Fourier pairs. We consider the transformation of functions in class L^p in chapter 8.

The ordinary Fourier pair,

$$|x|^{-1/2} \leftrightarrow |y|^{-1/2}$$

is an example in which neither function is in L^p for any p satisfying $1 \leqslant p \leqslant \infty$. Criteria for establishing the Fourier transformability of such functions are more involved, and one way of establishing these pairs is to show that they are the limit of suitably dominated sequences of ordinary Fourier pairs in L^p. We describe this in chapter 9. Note importantly that pointwise convergence of a sequence of Fourier pairs is not adequate in this context. For example, consider the ordinary pair $f_n \leftrightarrow F_n$ where, for any positive integer n,

$$f_n(x) = \begin{cases} n^2 x, & 0 \leqslant x \leqslant n^{-1} \\ 2n - n^2 x, & n^{-1} < x \leqslant 2/n \\ 0, & x<0 \text{ or } x>2/n. \end{cases} \tag{6.11}$$

The graph of f_n consists of a triangular hump just to the right of the origin, with unit area and height n. Since $f_n \in L$ we obtain F_n from (6.1), and for each n,

$$F_n(y) = \begin{cases} \exp\left(-\dfrac{2\pi i y}{n}\right)\left[\dfrac{\sin(\pi y/n)}{(\pi y/n)}\right]^2, & |y|>0 \\ 1, & y=0 \end{cases} \tag{6.12}$$

On considering the limit $n \to \infty$ we find that $f_n(x)$ tends pointwise to zero at *every* x, and that $F_n(y)$ tends pointwise to unity at *every* y. Nevertheless the functions $f(x)=0$ and $F(y)=1$ do not constitute a recognized or useful Fourier pair. In this example the sequence $\{f_n\}$ is more usefully regarded as tending to a delta function as $n \to \infty$, a rigorous basis for this approach being given in chapters 12 and 13.

6.4 Necessary conditions for transformability

If a function f belongs to an ordinary Fourier pair then the behaviour of $f(x)$ as $x \to \pm\infty$ is necessarily constrained, but the nature of the constraint is not simply described. It is certainly not necessary that $f(x) \to 0$ as $x \to \pm\infty$. Consider for instance the function $f(x)$ whose graph consists of positive rectangular pulses centred on $x = \pm 1, \pm 2, \ldots, \pm n, \ldots$, the pulse at $x = \pm n$ having height n and width n^{-3}, the function value in between pulses being zero. This function is $L(-\infty, \infty)$ and so has a continuous transform. Although the heights of the pulses tend to infinity as $n \to \pm\infty$, this is compensated by the pulses getting narrower. Indeed one can readily adapt the example to make the heights increase faster with n, provided the widths become correspondingly narrower.

It is necessary, however, that a suitably smoothed version of a function must tend to zero at infinity if the function belongs to an ordinary Fourier pair. More precisely, if f and its transform F are both locally integrable, then there will exist an ordinary function φ such that the convolution $f * \varphi$ is defined everywhere, and

$$\lim_{x \to \pm\infty} f * \varphi(x) = 0. \tag{6.13}$$

It follows, for example, that the functions $f(x) = 1$, $f(x) = \cos x$, $f(x) = x^2$ cannot belong to ordinary Fourier pairs, so that their transforms will only exist as generalized functions.

In view of the limit in (6.13) it might be thought that when f belongs to an ordinary Fourier pair then necessarily there will exist some positive number s (dependent perhaps on f) such that

$$\int_{-\infty}^{+\infty} \frac{|f(x)|}{(1+|x|)^s} \, \mathrm{d}x < \infty. \tag{6.14}$$

In fact this need not be the case. In chapter 9 we consider functions such as the following, defined at all x by the convergent series,

$$f(x) = \sum_{n=-\infty}^{+\infty} \exp(|n|) \cos(2\pi n x) \exp[-\pi(x-n)^2]. \tag{6.15}$$

This function belongs to an ordinary Fourier pair, and yet the integral in (6.14) diverges for every positive value of s. A feature of the function just defined is that growth in the peak values of $|f(x)|$ as $x \to \pm\infty$ is accompanied by increasingly rapid oscillations in the value of $f(x)$ between positive and negative values. If a function is non-negative then the situation is different, as the following theorem shows. Let f be a function belonging to an ordinary Fourier pair, and suppose that $f(x)$ is non-negative for almost all x, then it follows that the integral in (6.14) converges for each $s > 1$ (where s is not necessarily restricted to integer values).

Note that if f belongs to an ordinary Fourier pair then it will not necessarily follow that $|f|$ belongs to an ordinary pair. The function defined by equation (6.15) provides an example of this.

Finally, if an ordinary function tends to zero at $\pm\infty$ this by no means ensures that the function belongs to an ordinary Fourier pair, even though it is possible to construct functions belonging to ordinary Fourier pairs which decay arbitrarily slowly at infinity. We elaborate on these comments in chapter 8.

7

Fourier theorems for good functions

7.1 Introduction

A good function is defined as follows.

Definition Suppose a real or complex valued function $f(x)$ is defined for all real x and is everywhere infinitely differentiable, and suppose that each differential tends to zero as $x \to \pm\infty$ faster than any positive power of x^{-1}, or in other words suppose that for each positive integer m and each positive integer n,

$$\lim_{x \to \pm\infty} x^m f^{(n)}(x) = 0:$$

then we say that f is a *good function.* □

We will follow a common convention and represent the class of good functions by S. For example, the functions equal at all real x to $\exp(-x^2)$, $x\exp(-x^2)$, and $(1+x^2)^{-1}\exp(-x^2)$, respectively, are good functions, whilst the function $\exp(-|x|)$ is not good (because it is not differentiable at $x=0$) and the function $(1+x^2)^{-1}$ is not good (because it decays too slowly as $x \to \pm\infty$).

The good functions are important in Fourier theory because the inversion, convolution and differentiation theorems, as well as many others, apply in particularly simple forms with no problems of convergence. The two chief properties of good functions, those of rapid decay at infinity and of infinite differentiability, are related in that on Fourier transformation the one property transforms into the other: as a result the Fourier transform of a good function is also a good function.

Good functions are also important because of the role they play in the theory of generalized functions, and many of the results given in this chapter are necessary in understanding chapters 12–14. A good function of bounded support is a special type of good function that also plays an important part in the theory of generalized functions, and section 7.3 is devoted to the properties of these functions and their transforms. At a first reading section 7.3 could well be omitted.

Good functions have the following important properties. The sum (or difference) of two good functions is necessarily also a good function, and the product and convolution of two good functions is a good function. Differentiation of a good function yields another good function, and $x^n\varphi(x)$ is a good function for all non-negative integers n whenever φ is good. A good function is necessarily L^p for every p satisfying $1 \leqslant p \leqslant \infty$. The integral of a good function is not necessarily good however, and if φ is a good function then the function g defined for all x by

$$g(x) = \int_{-\infty}^{x} \varphi(u)\mathrm{d}u$$

will be good if, and only if, $\int \varphi = 0$.

Good functions are not only continuous but are also uniformly continuous on $(-\infty, \infty)$ and absolutely continuous on $(-\infty, \infty)$. However, a good function cannot necessarily be represented over every interval by a Taylor series expansion, as the example in (7.19) will show.

7.2 Inversion, differentiation and convolution theorems

We now state formally the fundamental theorems relating to the Fourier transformation of good function. First, the definition of the Fourier transform, and the inversion theorem.

***Theorem** 7.1* Let φ be a good function: then (i) a function Φ may be defined at all real y by

$$\Phi(y) = \int_{-\infty}^{\infty} \varphi(x) \exp(-2\pi\mathrm{i}xy)\mathrm{d}x, \tag{7.1}$$

and Φ will be a good function, and (ii), at each x,

$$\varphi(x) = \int_{-\infty}^{\infty} \Phi(y) \exp(2\pi\mathrm{i}xy)dy; \tag{7.2}$$

if g is any function equal a.e. to φ and if G is any function equal a.e. to Φ then we say g and G form a Fourier pair and write $g \leftrightarrow G$, G being a Fourier transform of g, and g being an inverse transform of G. □

The following are examples of Fourier pairs of good functions,

$$\exp(-\pi x^2) \leftrightarrow \exp(-\pi y^2) \tag{7.3}$$

$$2\cos(2\pi x)\exp(-\pi x^2) \leftrightarrow \exp[-\pi(x-1)^2] + \exp[-\pi(x+1)^2]. \tag{7.4}$$

***Theorem** 7.2* (Differentiation) Suppose $\varphi \leftrightarrow \Phi$ is a Fourier pair of good functions, and for $n = 1,2,3,\ldots$ *let* $\varphi^{(n)}$ and $\Phi^{(n)}$ be the nth derivatives of φ and Φ: then the following are Fourier pairs of good functions, for $n = 1,2,3,\ldots,$

$$(-2\pi \mathrm{i}x)^n \varphi(x) \leftrightarrow \Phi^{(n)}(y) \tag{7.5}$$

$$\varphi^{(n)}(x) \leftrightarrow (2\pi \mathrm{i}y)^n \Phi(y). \quad \square \tag{7.6}$$

For example on differentiating the left hand member of (7.3) and using (7.6) we obtain the following Fourier pair of good functions:

$$-x \exp(-\pi x^2) \leftrightarrow \mathrm{i}y \exp(-\pi y^2).$$

***Theorem** 7.3* (Integration) Let $\varphi \leftrightarrow \Phi$ be Fourier pair of good functions: then for each real x,

$$\int_0^x \varphi(u)\mathrm{d}u = \int_{-\infty}^{\infty} \Phi(y)\left[\frac{\exp(2\pi \mathrm{i}xy) - 1}{2\pi \mathrm{i}y}\right]\mathrm{d}y, \tag{7.7}$$

and

$$\varphi(x) = \frac{\mathrm{d}}{\mathrm{d}x}\int_{-\infty}^{\infty} \Phi(y)\left[\frac{\exp(2\pi \mathrm{i}xy) - 1}{2\pi \mathrm{i}y}\right]\mathrm{d}y. \quad \square \tag{7.8}$$

Note that the presence of $(2\pi \mathrm{i}y)$ in the denominator of the integrand in (7.7) and (7.8) causes no problem at $y = 0$, even when $\Phi(0)$ is non-zero, because the quantity in the square brace tends to the finite value x as $y \to 0$. The functions φ and Φ can be interchanged in (7.7) and (7.8) provided i is replaced by $-\mathrm{i}$ in each of the square braces.

***Theorem** 7.4* (Convolution and products) Suppose $\varphi \leftrightarrow \Phi$ and $\psi \leftrightarrow \Psi$ are Fourier pairs of good functions: then $\varphi * \psi$, $\Phi * \Psi$, $\varphi\psi$, and $\Phi\Psi$ are good functions and,

$$\varphi * \psi \leftrightarrow \Phi\Psi \tag{7.9}$$

and

$$\varphi\psi \leftrightarrow \Phi * \Psi. \quad \square \tag{7.10}$$

This theorem sometimes provides a convenient way of evaluating convolutions. For example, from (7.3) and (7.9),

$$[\exp(-\pi x^2)] * [exp(-\pi x^2)] \leftrightarrow \exp(-2\pi y^2); \tag{7.11}$$

but, from (7.3) and (6.6),

$$2^{1/2} \exp(-\pi x^2/2) \leftrightarrow \exp(-2\pi y^2) \tag{7.12}$$

so that on equating the left hand members of (7.11) and (7.12) we see that the convolution of a Gaussian with itself yields another Gaussian.

***Theorem** 7.5* (Parseval) Suppose $\varphi \leftrightarrow \Phi$ and $\psi \leftrightarrow \Psi$ are Fourier pairs of good functions: then it follows that

$$\int_{-\infty}^{+\infty} \varphi(x)\psi^*(x)\mathrm{d}x = \int_{-\infty}^{+\infty} \Phi(y)\Psi^*(y)\mathrm{d}y \tag{7.13}$$

and that

$$\int \varphi\psi = \int \Phi\Psi_{\mathrm{T}} = \int \Phi_{\mathrm{T}}\Psi \tag{7.14}$$

and

$$\int \varphi_{\mathrm{T}}\psi = \int \varphi\psi_{\mathrm{T}} = \int \Phi\Psi, \tag{7.15}$$

where $\varphi_{\mathrm{T}}, \psi_{\mathrm{T}}, \Phi_{\mathrm{T}}$ and Ψ_{T} are the transposes of φ, ψ, Φ and Ψ, see (6.2). □

Equation (7.13) is known as *Parseval's formula*, as also are the equivalent equations in (7.14) and (7.15). Parseval's formula arises in fact as a special case of the convolution theorem, equation (7.10), when $\Phi * \Psi(0)$ is obtained from $\varphi\psi$ by a Fourier integral. When $\varphi = \psi$ we have, of course, the important special case of Parseval's formula,

$$\int_{-\infty}^{\infty} |\varphi(x)|^2\mathrm{d}x = \int_{-\infty}^{\infty} |\Phi(y)|^2\mathrm{d}y. \tag{7.16}$$

Certain results follow more or less immediately from the above theorems. For instance, using a prime to represent the derivative of a function, for any good functions φ and ψ the functions $(\varphi * \psi)'$, $(\varphi' * \psi)$, and $(\varphi * \psi')$ are good and everywhere equal to each other and they have the good function $[2\pi i y\Phi(y)\Psi(y)]$ as their Fourier transform. The extension to higher derivatives follows, and for any non-negative integers, k, l and m, the functions $(\varphi^{(k)} * \psi^{(l)})^{(m)}$ and $(\varphi * \psi)^{(k+l+m)}$ will be everywhere equal when φ and ψ are good.

The cross correlation $\rho_{\varphi\psi}$ of two good functions φ and ψ is defined for all x by

$$\rho_{\varphi\psi}(x) = \int_{-\infty}^{\infty} \varphi^*(x')\psi(x+x')\mathrm{d}x',$$

and $\rho_{\varphi\psi}$ is also a good function; it then follows that

$$\rho_{\varphi\psi} \leftrightarrow \Phi^*\Psi. \tag{7.17}$$

In physical contexts the quantity $\Phi^*\Psi$ is often referred to as the cross energy spectrum of the functions φ and ψ. When φ and ψ are identical we have

$$\rho_{\varphi\varphi} \leftrightarrow |\Phi|^2, \tag{7.18}$$

so that $|\Phi|^2$, called the *energy spectrum* of φ, is the Fourier transform of $\rho_{\varphi\varphi}$, called the *autocorrelation* function of φ. These results are very closely related to the convolution theorem, since for all x

$$\rho_{\varphi\psi}(x) = \varphi_{\mathrm{T}}^{*} * \psi(x);$$

in other words the cross correlation function between φ and ψ is identical to the convolution between φ_{T}^{*} and ψ.

7.3 Good functions of bounded support

An ordinary function $f(x)$ is said to be of *bounded support* if there exists a number $a>0$ such that $f(x)=0$ whenever $|x| \geqslant a$; the class of all good functions of bounded support is often denoted by D in contrast to the class of all good functions denoted by S. As an example we have $\rho \in D$, where

$$\rho(x) = \begin{cases} \exp\{-1/(1-x^2)\}, & |x|<1 \\ 0, & |x| \geqslant 1. \end{cases} \tag{7.19}$$

Note particularly that ρ is infinitely differentiable at $x = \pm 1$, as it must be in order to be good. The function ρ shows that a good function does not necessarily obey a Taylor series expansion over every interval, since a Taylor expansion based on the various derivatives of ρ for any point having $|x|>1$ would lead to zero value at all x.

It can be shown that the Fourier transform of a function in D cannot also be in D except in the trivial case that both functions are zero almost everywhere, and it is common to represent the class of all good functions that are Fourier transforms of functions in D by the letter Z. The classes D and Z are subsets of S.

For a good function Φ to be in class Z it is necessary, but not sufficient, that it should obey a Taylor series expansion so that for any pair of real numbers y and y_1

$$\Phi(y_1) = \Phi(y) + \Phi'(y)(y_1 - y) + \frac{\Phi''(y)(y_1-y)^2}{2!} + \cdots$$
$$+ \frac{\Phi^{(k)}(y)(y_1-y)^k}{k!} + \cdots.$$

If functions φ and ψ are in D, then so also are the various functions $\varphi+\psi$, $\varphi\psi$, $\varphi * \psi$, φ', $\varphi^{(k)}$, $x^k\varphi(x)$, for $k=0, 1, 2, \ldots$.

Many of the special properties of functions in Z are best displayed when the domain of definition is extended beyond the real line to cover the whole of the complex plane. To do this we introduce the complex variable $z = y + i\alpha$, where y and α are real, and define the complex Fourier transform Φ_{c} of some $\varphi \in D$ as follows. Given any function $\varphi \in D$, there exists a complex valued function Φ_{c}, with domain the complex plane, such that for every complex z,

$$\Phi_c(z)=\int_{-\infty}^{\infty} \varphi(x)\exp(-2\pi i x z)dx; \qquad (7.20)$$

we call Φ_c the complex Fourier transform of φ and write

$$\varphi(x)\leftrightarrow\Phi_c(z).$$

It is clear that the Fourier transform Φ of φ can be obtained from the complex Fourier transform Φ_c, since for $\alpha=0$, $z=y$ we have

$$\Phi_c(y)=\Phi(y).$$

It is also true that Φ_c is uniquely determined by Φ, one way of obtaining Φ_c from Φ being to inversely transform Φ to give φ using (7.2), and then to use equation (7.20). We call Φ_c the analytic extension of Φ and in a moment will show other ways of obtaining Φ_c from Φ. It will be convenient to use the symbol Z_c to represent the class of all complex Fourier transforms of functions in D.

The Fourier inversion theorem for functions in Z_c can be expressed as follows.

***Theorem* 7.6** Suppose $\varphi\leftrightarrow\Phi_c$ where φ is in class D and Φ_c is in class Z_c: then for each real x and for arbitrary real α,

$$\varphi(x)=\exp(-2\pi\alpha x)\int_{-\infty}^{\infty} \Phi_c(y+i\alpha)\exp(2\pi i x y)dy \qquad (7.21)$$

$$=\int_{i\alpha-\infty}^{i\alpha+\infty} \Phi_c(z)\exp(2\pi i x z)dz. \quad \square \qquad (7.22)$$

The expression on the right hand side of (7.22) is just another way of writing the expression on the right hand side of (7.21), and the integral in (7.22) represents integration along a line in the complex plane parallel to the real axis. This inversion theorem follows readily from that in theorem 7.1 when it is noted that $\Phi_c(y+i\alpha)$, treated as a function of y at fixed α, is the ordinary Fourier transform of $\varphi(x)\exp(2\pi\alpha x)$.

We now quote some further properties of functions in Z and Z_c, moving towards a complete characterization. It can be shown that functions in Z_c are necessarily *entire functions*, where by definition an entire function is a complex valued function which is defined and differentiable everywhere on the complex plane. The special properties of functions of a complex variable are described in detail, for instance, in Apostol (1974) and Churchill (1960). For our present purposes, however, we will need to draw only on the following facts concerning the differentiation and Taylor series expansion of such functions.

Differentiation of a function of the complex variable z is defined by a limiting process analogous to that used for a function of a real variable, with the additional requirement that the limit must be independent of the direction chosen for the infinitesimal element δz in the complex plane. More precisely, given a real or complex valued function $G(z)$, defined on a region of the complex plane that includes all points within some circle of finite radius centred on a point z_0, we say G is differentiable at z_0 and has the derivative $G'(z_0)$ if for each $\varepsilon > 0$ there exists a $\delta > 0$, dependent on ε, such that whenever $|z - z_0| < \delta$ then

$$\left|\frac{G(z) - G(z_0)}{z - z_0} - G'(z_0)\right| < \varepsilon;$$

when these conditions are met we write

$$\lim_{z \to z_0} \left\{\frac{G(z) - G(z_0)}{z - z_0}\right\} = G'(z_0).$$

It can be shown that an entire function is necessarily continuous at all points on the complex plane, and also (surprisingly perhaps) that the function obtained on differentiating an entire function is also entire, so that an entire function is infinitely differentiable everywhere.

It can also be shown that the values of an entire function at any two points z and z_1 on the complex plane are related by a Taylor series thus:

$$G(z_1) = G(z) + G'(z)(z_1 - z) + \frac{G''(z)(z_1 - z)^2}{2!} + \cdots + \frac{G^{(k)}(z)(z_1 - z)^k}{k!} + \cdots. \tag{7.23}$$

This series shows another way of analytically extending a function $\varphi \in Z$ to the corresponding $\Phi_c \in Z_c$. Indeed a knowledge of the function values of some $\Phi \in Z$ over any finite interval of the real axis completely determines both $\Phi(y)$ and $\Phi_c(z)$ for all y and z.

Not every entire function, however, is necessarily in Z_c. For instance it is known that any polynomial,

$$G(z) = a_0 + a_1 z + a_2 z^2 + \cdots + a_n z^n \qquad (n = 0, 1, 2, \ldots)$$

is necessarily an entire function; G, however, will not be in Z_c except in the trivial case that all the coefficients $a_0, a_1, \ldots, a_n$ are zero. Likewise the exponential function, defined for any complex $z = y + \mathrm{i}\alpha$ by

$$\exp z = (\cos \alpha + \mathrm{i} \sin \alpha) \exp y,$$

is an entire function, but is not in Z_c.

We are now in a position to quote a direct characterization of function in Z_c.

Theorem 7.7 A function Φ_c will be the complex Fourier transform of some good function φ for which $\varphi(x)=0$ whenever $|x|>b,\ b>0$, if and only if the following conditions are both satisfied:

(i) Φ_c is an entire function
(ii) for each non-negative integer k there exists a positive number C_k such that for all values of $z=y+i\alpha$

$$|z^k\Phi_c(z)|\leqslant C_k\exp(b|\alpha|).\quad\square \tag{7.24}$$

There exists yet another way of extending a function $\Phi\in Z$ to the corresponding $\Phi_c\in Z_c$, as follows. Suppose that $\varphi\in D$ and that $\varphi(x)=0$ whenever $|x|>b,\ b>0$; if Φ and Φ_c are the Fourier transform and complex Fourier transform, respectively, of φ, defined as in (7.1) and (7.20), then

$$\frac{1}{2\pi i}\int_{-\infty}^{+\infty}\Phi(y)\frac{\exp[-2\pi ib(z-y)]}{z-y}\,dy=\begin{cases}-\Phi_c(z), & \operatorname{Im} z>0\\ 0, & \operatorname{Im} z<0\end{cases} \tag{7.25}$$

and

$$\frac{1}{2\pi i}\int_{-\infty}^{+\infty}\Phi(y)\frac{\exp[+2\pi ib(z-y)]}{z-y}\,dy=\begin{cases}0, & \operatorname{Im} z>0\\ \Phi_c(z), & \operatorname{Im} z<0.\end{cases} \tag{7.26}$$

We end by showing how several results quoted already in equations (6.3)–(6.7) and in section 7.2 can be extended to apply to the complex Fourier transformation of functions in D. We first need to attribute meaning to $\Phi_c*\Psi_c$, the convolution of two functions in Z_c. Suppose that $\Phi\in Z$ and $\Psi\in Z$, and that their respective extensions in Z_c are Φ_c and Ψ_c: it follows that there exists a function $\Phi_c*\Psi_c$ in Z_c defined for every value of the complex variable z by

$$\begin{aligned}\Phi_c*\Psi_c(z)&=\int_{-\infty}^{+\infty}\Phi_c(y+i\alpha)\Psi_c(z-y-i\alpha)\,dy\\ &=\int_{i\alpha-\infty}^{i\alpha+\infty}\Phi_c(z')\Psi_c(z-z')\,dz'\end{aligned}$$

where y is a real variable, z' is a complex variable, and the value of the integral is independent of the value of the real number α; it also follows that the convolution $\Phi_c*\Psi_c$, so defined, is equal to the extension into Z_c of the convolution $\Phi*\Psi$ in Z.

The theorems on linearity, shifting, scaling, complex conjugation, transposition, differentiation, products and convolution now take the following forms. Suppose that $\varphi\in D$ and $\psi\in D$ have $\Phi_c\in Z_c$ and $\Phi_c\in Z_c$ as their respective complex Fourier transforms; then for each complex number β, each complex number γ, each real a, and each non-zero real number b, the following Fourier pairs exist where in each case the expression on the left (with variable x) represents a function in D whilst the expression on the right represents the complex Fourier transform (with complex variable z):

$$(\beta\varphi+\gamma\psi)\leftrightarrow(\beta\Phi_c+\gamma\Psi_c)$$

$$\varphi(x-a)\leftrightarrow\exp(-2\pi iaz)\Phi_c(z)$$

$$\exp(2\pi i\beta x)\varphi(x)\leftrightarrow\Phi_c(z-\beta)$$

$$\varphi^*(x)\leftrightarrow\Phi_c^*(-z^*)$$

$$\varphi(-x)\leftrightarrow\Phi_c(-z)$$

$$\varphi^*(-x)\leftrightarrow\Phi_c^*(z^*)$$

$$\varphi^{(k)}(x)\leftrightarrow(2\pi iz)^k\Phi_c(z)$$

$$(-2\pi ix)^k\varphi(x)\leftrightarrow\Phi_c^{(k)}(z)$$

$$\varphi*\psi\leftrightarrow\Phi_c\Psi_c$$

$$\varphi\psi\leftrightarrow\Phi_c*\Psi_c.$$

Finally we have a version of the Parseval formula. Given $\varphi\in D$ and $\psi\in D$ with their respective complex Fourier transforms $\Phi_c\in Z_c$ and $\Psi_c\in Z_c$, then the following equalities hold where y is a real variable and α is an arbitrary real number:

$$\int_{-\infty}^{+\infty}\varphi(x)\psi(-x)\mathrm{d}x=\int_{-\infty}^{+\infty}\Phi_c(y+i\alpha)\Psi_c(y+i\alpha)\mathrm{d}y$$

$$\int_{-\infty}^{+\infty}\varphi(x)\psi^*(x)\mathrm{d}x=\int_{-\infty}^{+\infty}\Phi_c(y+i\alpha)\Psi_c^*(y-i\alpha)\mathrm{d}y$$

$$\int_{-\infty}^{+\infty}\varphi(x)\psi(x)\mathrm{d}x=\int_{-\infty}^{+\infty}\Phi_c(y+i\alpha)\Psi_c(-y-i\alpha)\mathrm{d}y.$$

8

Fourier theorems in L^p

8.1 Basic theorems and definitions

When a function $f(x)$ belongs to $L^p(-\infty, \infty)$, for some p satisfying $1 \leqslant p \leqslant 2$, it is possible to assign to f a locally integrable Fourier transform $F(y)$ in a useful manner, and various formulae exist relating F to f and vice versa. The cases $p=1$ and $p=2$ are particularly common in physical applications, and we consider first the case $p=1$.

Theorem 8.1 Suppose f belongs to $L(-\infty, \infty)$: then at all real y a function $F(y)$ may be defined by

$$F(y) = \int_{-\infty}^{\infty} f(x) \exp(-2\pi i x y) dx, \tag{8.1}$$

and it will follow that:

(i) F is bounded on $(-\infty, \infty)$ and at each y,

$$|F(y)| \leqslant \int_{-\infty}^{\infty} |f(x)| dx,$$

(ii) F is everywhere continuous and indeed uniformly continuous on $(-\infty, \infty)$,

(iii) $F(y)$ tends to zero as $y \to \pm\infty$. □

The function F, together with any function equal a.e. to F, is called the Fourier transform of f, and in Table 8.1 numbers (8.3) to (8.7) provide examples of such transforms. The remaining examples (8.8) and (8.9) will be discussed later. In the table $r(x)$ is the rectangular function

$$r(x) = \begin{cases} 1, & |x| < 1 \\ 0, & |x| > 1 \\ \tfrac{1}{2}, & |x| = 1. \end{cases} \tag{8.2}$$

Table 8.1. *Table of transforms in L^p*

$f(x)$	$F(y)$	
$\frac{1}{2}\exp(-\|x\|)$	$(1+4\pi^2y^2)^{-1}$	(8.3)
$(1-\|x\|)r(x)$	$\left(\frac{\sin \pi y}{\pi y}\right)^2$ [$=0$ at $y=0$]	(8.4)
$r(x)$	$\frac{\sin 2\pi y}{\pi y}$ [$=2$ at $y=0$]	(8.5)
$\exp(-\|x\|)H(x)$	$(1+2\pi i y)^{-1}$	(8.6)
$\|x\|^{-1/2}\exp(-2\pi\|x\|)$	$\left[\frac{1+(1+y^2)^{1/2}}{1+y^2}\right]^{1/2}$	(8.7)
$\frac{\sin x}{\|x\|}$	$-i(\operatorname{sgn} y)\ln\left\|\frac{\|2\pi y\|+1}{\|2\pi y\|-1}\right\|$	(8.8)
$\frac{[1-r(x)]\operatorname{sgn} x}{(x^2-1)^{1/2}}$	$-\pi i(\operatorname{sgn} y)J_0(2\pi y)$	(8.9)

$H(x)$ is the Heaviside step function, (4.7), and $\operatorname{sgn} x$ is the signum function (5.3).

We notice from Table 8.1 that when $f \in L$, the transform F may or may not be in L, since $F \in L$ in (8.3) and (8.4) but $F \notin L$ in (8.5)–(8.7). When F is integrable we have the following straightforward inversion theorem.

Theorem 8.2 Suppose f is in class L, let F be defined by (8.1) at all y, and suppose that $F \in L$: then it follows that f is equal a.e. to a continuous function and, supposing for simplicity that f is continuous, it follows that at all x,

$$f(x) = \int_{-\infty}^{\infty} F(y)\exp(+2\pi i x y)\mathrm{d}y. \quad \square$$

When the transform F is not in L then the above integral will diverge at all x, and the inversion theorem will need modifying. One way is as follows.

Theorem 8.3 Suppose $f \in L$ has the transform F defined everywhere by (8.1): then it follows that

$$\lim_{\lambda\to\infty} \int_{-\infty}^{\infty} F(y)\exp(-|y|/\lambda)\exp(2\pi i x y)\mathrm{d}y \tag{8.10}$$

converges in the following senses:

(i) to $f(x)$ on the Lebesgue set of f and so at a.a.x,
(ii) to $f(x)$ at each point of continuity of f,

(iii) to $\frac{1}{2}[f(x^+)+f(x^-)]$ at each value of x for which these limits exist,
(iv) to the Lebesgue value $f_L(x)$ at each Lebesgue point of f. □

The convergence factor, $\exp(-|y|/\lambda)$, appearing in theorem 8.2 is one of several alternatives which we discuss in section 8.2.

We now consider the Fourier transformation of a function that is L^p for some p satisfying $1<p\leqslant 2$. Two approaches are common, one based on mean convergence and the other on the use of a convergence factor: a definition based on (8.1) is no longer available since we do not assume that f is necessarily in class L. We consider mean convergence first.

Theorem 8.4 Suppose that f is in $L^p(-\infty,\infty)$ for some p satisfying $1<p\leqslant 2$: then there will exist a function F in L^q, where $p^{-1}+q^{-1}=1$, such that

$$F(y)=\underset{\lambda\to\infty}{\text{l.i.m.}}(q)\int_{-\lambda}^{\lambda} f(x)\exp(-2\pi i x y)\mathrm{d}x, \tag{8.11}$$

and it will follow that

$$f(x)=\underset{\lambda\to\infty}{\text{l.i.m.}}(p)\int_{-\lambda}^{\lambda} F(y)\exp(2\pi i x y)\mathrm{d}y. \quad \square \tag{8.12}$$

Transforms (8.8) and (8.9) provide examples of functions satisfying the conditions of the theorem. In (8.8) $f\in L^p$ for all p in the range $1<p\leqslant\infty$, whilst in (8.9) $f\in L^p$ for all $p\in(1,2)$ but not for other values of p.

We note that theorem 8.4 defines $F(y)$ uniquely apart from its value on a null set, and also that when a function is L^p for two or more values of p in the range $1\leqslant p\leqslant 2$ then the values of F which satisfy theorems 8.1 and 8.4 are equal a.e. The case $p=2$ is unique, for in that case $p=q$ so that square integrable functions transform into square integrable functions.

The use of an exponential convergence factor leads to the following.

Theorem 8.5 Suppose that f belongs to L^p for some p satisfying $1<p\leqslant 2$: then at a.a.y we may define a function F by

$$F(y)=\lim_{\lambda\to\infty}\int_{-\infty}^{\infty} f(x)\exp(-|x|/\lambda)\exp(-2\pi i x y)\mathrm{d}x, \tag{8.13}$$

and it will follow that

$$\lim_{\lambda\to\infty}\int_{-\infty}^{\infty} F(y)\exp(-|y|/\lambda)\exp(2\pi i x y)\mathrm{d}y \tag{8.14}$$

will converge to $f(x)$ at a.a.x and in each of the senses (i)–(iv) given in theorem 8.3. □

Once again the function F so defined is essentially equal to that given by theorem 8.4 or, when $f\in L$, by theorem 8.1.

The following theorem provides yet another variant.

Theorem 8.6 Suppose $f \in L^p$ for some $p \in (1,2]$: then at a.a.y a function F may be defined by

$$F(y) = \lim_{\lambda \to \infty} \int_{-\lambda}^{\lambda} f(x) \exp(-2\pi i x y) dx \tag{8.15}$$

and it will follow that at a.a.x,

$$f(x) = \lim_{\lambda \to \infty} \int_{-\lambda}^{\lambda} F(y) \exp(2\pi i x y) dy. \quad \square \tag{8.16}$$

Although this theorem seems to provide the simplest definition of the transform when $p \in (1,2]$, the proof is more involved than is the case for theorems 8.4 and 8.5, and indeed the validity of (8.15) and (8.16) remained an open question until the work of Carleson (1966) and Hunt (1968). Note also that if $f \in L$, and F is defined by (8.1) or (8.15), then it does not necessarily follow that (8.16) will converge for *any* x: we discuss this further in section 8.2.

We have so far placed the integrability condition on f, but an analogous set of results can be obtained if the condition is placed on F. With this in mind, and with the aim of formulating a single definition applicable whenever $p \in [1,2]$, we now state the following definition of a Fourier pair in L^p.

Definition Suppose two functions f and F are related as follows, for a.a.y and a.a.x, respectively:

$$F(y) = \lim_{\lambda \to \infty} \int_{-\infty}^{\infty} f(x) \exp(-|x|/\lambda) \exp(-2\pi i x y) dx$$

$$f(x) = \lim_{\lambda \to \infty} \int_{-\infty}^{\infty} F(y) \exp(-|y|/\lambda) \exp(2\pi i x y) dy,$$

and suppose that at least one of the functions is L^p for some $p \in [1,2]$: then it will follow that the other function is L^q, where $p^{-1} + q^{-1} = 1$ (∞^{-1} and zero being interchangeable); when these conditions are met we say that f and F are a Fourier pair in L^p, we write $f \leftrightarrow F$, we call F the Fourier transform of f and call f the inverse Fourier transform of F. $\square$

8.2 More inversion theorems in L^p

We now elaborate and extend the theorems already given in section 8.1. We will start with equations based on a formula using no convergence factor and then consider the use of convergence factors; we follow this by a consideration of formulae based on limits of the type $\lim_{\lambda \to \infty} \int_{-\lambda}^{\lambda}$ and then turn to convergence in the mean and certain miscellaneous theorems.

In order that a Fourier pair $f \leftrightarrow F$ shall satisfy each of the following, at all y and x, respectively:

$$F(y) = \int_{-\infty}^{\infty} f(x)\exp(-2\pi ixy)\mathrm{d}x \tag{8.17}$$

$$f(x) = \int_{-\infty}^{\infty} F(y)\exp(2\pi ixy)\mathrm{d}y \tag{8.18}$$

it is necessary that both f and F shall be in class L, as in theorem 8.2. It is natural to enquire whether this theorem can be adapted so that all the conditions are placed on f, instead of being spread between f and F. One needs a condition which will ensure that a function is the transform (or inverse transform) of some integrable function. It is notable that no simple condition that is both necessary and sufficient for this is known. In particular, the fact that a function F is continuous and tends to zero at infinity (see theorem 8.1) is not sufficient to ensure that F is the transform of some integrable function.

However, there are certain simple conditions which are sufficient to ensure that a function is the transform (or inverse transform) of a function in L; the following theorems are based, respectively, on convexity and on differentiability.

Theorem 8.7 Suppose a continuous function f is defined, real and bounded on $(-\infty, \infty)$ and that $f(x) = f(-x)$ at all x; suppose also that (i) for $x \geqslant 0$ $\lim_{x\to\infty} f(x) = 0$, and that (ii) $f(x)$ is convex downwards for $x \geqslant 0$: then f belongs to a Fourier pair in which the other function is $L(-\infty, \infty)$ and is nowhere negative. □

The conditions (i) and (ii) on f mean, more explicitly, that when $0 < x_1 < x_2$ then $f(x_1) \geqslant f(x_2)$, the equality sign applying only when $f(x_1) = 0$, and that when $0 \leqslant x_1 < x_2 < x_3$ then

$$\frac{f(x_2) - f(x_3)}{x_3 - x_2} \leqslant \frac{f(x_1) - f(x_3)}{x_3 - x_1}, \tag{8.19}$$

so that on $x \geqslant 0$ a chord never goes below the graph of $f(x)$. For example, the exponential and 'triangle' functions, $f(x)$ in (8.3) and (8.4), satisfy the conditions of theorem 8.7. Likewise the function F defined everywhere by

$$F(y) = [\ln(2 + |y|)]^{-1} \tag{8.20}$$

is the transform of some function $f \in L$, even though F is not in L^p for any finite $p \geqslant 1$.

Theorem 8.8 Suppose $f \in L^p$ for some $p \in [1, 2]$ and that f is absolutely continuous with a derivative f' that is L^r for some $r \in (1, 2]$: then it will follow that the transform and inverse transform of f are each $L(-\infty, \infty)$. □

For example, the functions $f(x)$ in (8.3) and (8.4) satisfy the conditions of this theorem, as also do the functions $F(y)$ in (8.3)–(8.7).

On combining the two results above we can construct the following inversion theorem.

Theorem 8.9 Suppose a continuous function $f \in L$ can be expressed as the sum of two functions satisfying the conditions of theorems 8.7 (convexity) and 8.8 (differentiability), respectively: then f will possess a continuous transform $F \in L$ and (8.17) and (8.18) will be valid at all y and x, respectively. □

We now broaden the discussion to include convergence factors, and show that the factor $K(y) = \exp(-|y|)$, used already in theorems 8.3 and 8.5, is but one amongst an infinity of suitable choices. We start by considering a convergence function K satisfying the conditions:

(C1) K is $L(-\infty, \infty)$ and defined on $(-\infty, \infty)$,
(C2) K is real valued and $K(y) = K(-y)$ at all y,
(C3) K is everywhere continouous and $K(0) = 1$,
(C4) K is the Fourier transform of a function k such that for some $s > 1$ and some $A > 0$ and for all x,

$$|k(x)| < A(1 + |x|)^{-s}.$$

For example, the functions $K(y) = \exp(-|y|)$, $\exp(-y^2)$, and $(1 + y^2)^{-1}$ each satisfy the conditions (C1)–(C4). So also does the triangle function $K(y) = (1 - |y|)$ $(=0$ for $|y| > 1)$; this is often called the Fejér convergence factor or kernel. The rectangle function (8.2) however fails to satisfy (C3) or (C4). We now have the following very powerful set of results.

Theorem 8.10 Suppose $f \in L^p$ for some $p \in [1, 2]$ and that K is a function satisfying all of conditions (C1) to (C4) above: then it follows that f has a transform $F \in L^q$, where $p^{-1} + q^{-1} = 1$ $(\infty^{-1} \equiv 0)$ and that

$$\lim_{\lambda \to \infty} \int_{-\infty}^{\infty} f(x) K(x/\lambda) \exp(-2\pi i x y) \mathrm{d}x \tag{8.21}$$

will converge in the following senses:

(i) to $F(y)$ on the Lebesgue set of F, and so almost everywhere,
(ii) to $F(y)$ at each point of continuity of F,
(iii) to $\frac{1}{2}[F(y^+) + F(y^-)]$ at each point where these limits exist, and to $\lim_{\varepsilon \to 0} \frac{1}{2}[F(y + \varepsilon) + F(y - \varepsilon)]$ at each point where this limit exists,

(iv) to the Lebesgue value $F_L(y)$ at each Lebesgue point of F,
(v) uniformly to F on an interval $[c, d]$ when F is continuous on (a, b) where $a<c<d<b$,
(vi) uniformly to F on $(-\infty, \infty)$ when $f \in L$ and F is continuous,
(vii) to $F(y)$ as a l.i.m.$_{\lambda\to\infty}(q)$ whenever $1<p\leqslant 2$;

likewise it will follow also that

$$\lim_{\lambda\to\infty}\int_{-\infty}^{\infty} F(y)K(y/\lambda)\exp(2\pi ixy)dy \tag{8.22}$$

will converge to $f(x)$ in senses (i) to (v) above (with F replaced by f) and also as a l.i.m.$_{\lambda\to\infty}(p)$. □

When the Fejér convergence factor is used, then (8.21) and (8.22) take the commonly used forms,

$$\lim_{\lambda\to\infty}\int_{-\lambda}^{\lambda} f(x)\left(1-\frac{|x|}{\lambda}\right)\exp(-2\pi ixy)dx \tag{8.23a}$$

and

$$\lim_{\lambda\to\infty}\int_{-\lambda}^{\lambda} F(y)\left(1-\frac{|y|}{\lambda}\right)\exp(2\pi ixy)dy. \tag{8.23b}$$

Theorem 8.10 may be modified by omitting the condition (C2) on the convergence function K provided that the limit values in part (iii) are replaced by the limit $AF(y^+)+BF(y^-)$, when it exists, where $A=\int_{-\infty}^{0} k$ and $B=1-A$, and provided also the limit value $F_L(y)$ in part (iv) is replaced by $AF_{L+}(y)+BF_{L-}(y)$ at points where these right and left hand Lebesgue values exist. The function K may now be complex valued and need not be an even function.

If the conditions on the convergence factor K are relaxed then the theorem is weakened: in the following version convergence at a simple step continuity or at a point of continuity is however retained.

Theorem 8.11 Suppose $f\in L^p$ for some $p\in[1,2]$ and that K is a function satisfying conditions (C1) to (C3) above and that also K is the transform of a continuous function $k\in L$ such that $\lim_{x\to\pm\infty} xk(x)=0$: then if F is a Fourier transform of f it will follow that

$$\lim_{\lambda\to\infty}\int_{-\infty}^{\infty} f(x)K(x/\lambda)\exp(-2\pi ixy)dx$$

will converge to $F(y)$ in senses (ii), (iii), (v), (vi) and (vii) of theorem 8.10, and likewise

$$\lim_{\lambda\to\infty}\int_{-\infty}^{\infty} F(y)K(y/\lambda)\exp(2\pi ixy)dy$$

will converge to $f(x)$ in senses (ii), (iii), (v) and (vi) of theorem 8.10 (with F replaced by f) and also as a $\text{l.i.m.}_{\lambda\to\infty}(p)$. □

We now discuss Fourier inversion theorems based on the following limits:

$$\lim_{\lambda\to\infty}\int_{-\lambda}^{\lambda} f(x)\exp(-2\pi i x y)\mathrm{d}x \tag{8.24}$$

and

$$\lim_{\lambda\to\infty}\int_{-\lambda}^{\lambda} F(y)\exp(2\pi i x y)\mathrm{d}y. \tag{8.25}$$

If $f\in L$ then (8.24) converges uniformly on $(-\infty,\infty)$ to a continuous limit function, $F(y)$; however, it is then possible that (8.25) may diverge at *all* x. A function $f\in L$ for which (8.25) diverges at all x is constructed by forming the product of the periodic Kolmogoroff function (see section 5.7) with either $\exp(-x^2)$ or $\exp(-|x|)$. Even when $f\in L$ and is everywhere continuous it is still possible that the inversion integral (8.25) may diverge at a non-denumerable infinity of points on each finite range of x; such a function arises when one of the periodic functions referred to in section 5.5 is multiplied by either $\exp(-x^2)$ or $\exp(-|x|)$.

The existence or otherwise of a Dirichlet value of a function at a particular point, (5.8), provides a crucial test for convergence or otherwise of (8.24) and (8.25) as follows.

Theorem 8.12 Suppose $f\in L^p$ for some $p\in[1,2]$ and suppose $f\leftrightarrow F$: then it follows that

(i) (8.24) will converge to $F(y)$ at a.a.y,
(ii) (8.24) will converge at a particular value y if, and only if, y is a Dirichlet point of F in which case the limit will equal $F_D(y)$,
(iii) (8.25) will converge at a particular value of x if, and only if, x is a Dirichlet point of f in which case the limit will equal $f_D(x)$,
(iv) (8.25) will converge a.e. to $f(x)$ on any finite or infinite interval (a, b) on which f is either continuous, bounded, or $L^r(a,b)$ for some $r>1$. □

Those parts of the above theorem dealing with convergence a.e. stem from the theorems of Carleson and Hunt referred to in section 5.6. We remind the reader that for a point to be a Dirichlet point of a function it is sufficient that the function is differentiable at the point or that the function is of bounded variation around the point: for more general conditions we refer back to the conditions of Jordan, Dini and others described in sections 5.8–5.11.

Uniform convergence over an interval is guaranteed if a condition based on bounded variation is used.

Theorem 8.13 Suppose $f \leftrightarrow F$ and that at least one of these functions is L^p for some $p \in [1, 2]$: then

$$\lim_{\lambda \to \infty} \int_{-\lambda}^{\lambda} f(x) \exp(-2\pi i x y) \mathrm{d}x \tag{8.26}$$

will converge uniformly to $F(y)$ on an interval $[a, b]$ whenever there exists an interval $[a-\varepsilon, b+\varepsilon]$, $\varepsilon > 0$, on which F is continuous and of bounded variation; likewise

$$\lim_{\lambda \to \infty} \int_{\lambda}^{\lambda} F(y) \exp(2\pi i x y) \mathrm{d}y \tag{8.27}$$

will converge uniformly to $f(x)$ on an interval $[a, b]$ whenever there exists an interval $[a-\varepsilon, b+\varepsilon]$, $\varepsilon > 0$, on which f is continuous and of bounded variation. □

The convergence of (8.27) on either side of a step discontinuity in $f(x)$ is such that an 'overshoot' occurs as $\lambda \to \infty$; this is known as the Gibbs phenomenon and it is discussed further in section 10.2.

When f is square integrable the approximation to f given by (8.27) for finite λ is the best possible in the following mean square sense.

Theorem 8.14 Suppose $f \in L^2$ has Fourier transform F and that G is some function in L^2, and let f_λ and g_λ be defined for $\lambda > 0$ by

$$f_\lambda = \int_{-\lambda}^{\lambda} F(y) \exp(2\pi i x y) \mathrm{d}y$$

$$g_\lambda = \int_{-\lambda}^{\lambda} G(y) \exp(2\pi i x y) \mathrm{d}y:$$

then it follows that for each λ,

$$\int_{-\infty}^{\infty} |f_\lambda(x) - f(x)|^2 \mathrm{d}x \leqslant \int_{-\infty}^{\infty} |g_\lambda(x) - f(x)|^2 \mathrm{d}x, \tag{8.28}$$

and the equality sign in (8.28) is valid if, and only if, $F = G$ a.e. on $(-\lambda, \lambda)$. □

We now consider some results based on mean convergence.

Theorem 8.15 Consider a sequence $\{f_n\}_{n=1}^{\infty}$ of functions each $L(-\infty, \infty)$ and suppose the sequence converges in L to a limit function $f \in L$: then the corresponding Fourier transforms converge uniformly to the transform of the limit in the sense that

$$\lim_{n\to\infty}\int_{-\infty}^{\infty} f_n(x)\,e^{-2\pi ixy}\,dx=\int_{-\infty}^{\infty} f(x)\,e^{-2\pi ixy}\,dx$$

converges uniformly on $-\infty<y<\infty$. □

Clearly theorem 8.15 can be modified to apply to a sequence $\{F_n\}$ converging in L, in which case the inverse transforms f_n will converge uniformly.

Theorem 8.15 has its counterparts for values of p other than unity as follows.

Theorem 8.16 Suppose that for some p satisfying $1<p\leqslant 2$,

$$\underset{n\to\infty}{\text{l.i.m.}}\,(p)f_n=f \tag{8.29}$$

where the functions f_n, $n=1,2,3,\ldots$, and f are in L^p: then it follows that

$$\underset{n\to\infty}{\text{l.i.m.}}(q)F_n=F \tag{8.30}$$

where $p^{-1}+q^{-1}=1$, $f_n\leftrightarrow F_n$ at each n, and $f\leftrightarrow F$. □

This theorem can likewise be applied to a sequence $\{F_n\}$ converging in L^p to F, $p\in(1,2]$, in which case the inverse transforms f_n converge in mean to the limit f, where $f\leftrightarrow F$.

The following provides yet another pair of equations relating the functions in a Fourier pair.

Theorem 8.17 Suppose $f\in L^p$ for some $p\in(1,2]$: then f will possess a transform F and at a.a.x and a.a.y, respectively, it will follow that:

$$F(y)=\frac{d}{dy}\int_{-\infty}^{\infty} f(x)\left[\frac{\exp(-2\pi ixy)-1}{-2\pi ix}\right]dx \tag{8.31}$$

$$f(x)=\frac{d}{dy}\int_{-\infty}^{\infty} F(y)\left[\frac{\exp(2\pi ixy)-1}{2\pi iy}\right]dy. \;\square \tag{8.32}$$

A slight modification of the above theorem allows the case $p=1$ to be included, thus providing an inversion theorem applicable pointwise a.e. to the Kolmogoroff function.

Theorem 8.18 Suppose $f\in L$: then f will possess a continuous transform F defined at all y by (8.1), and it will follow that at a.a.x,

$$f(x)=\frac{d}{dx}\left\{\lim_{\lambda\to\infty}\int_{-\lambda}^{\lambda} F(y)\left[\frac{\exp(2\pi ixy)-1}{2\pi iy}\right]dy\right\},$$

where the limit as $\lambda\to\infty$ converges at all x. □

Before concluding we make some further remarks on the properties of functions belonging to ordinary Fourier pairs in L^p. The fact that a continuous function decays to zero at infinity does not ensure that it is the transform of some integrable function. So likewise the fact that a function belongs to L^q, for some q satisfying $2<q<\infty$, does not ensure that it is the transform of some function in L^p, where $p^{-1}+q^{-1}=1$; examples are given in Titchmarsh (1962). However, there do exist functions in $L(-\infty,\infty)$ whose transforms decay arbitrarily slowly at infinity; this is stated more precisely as follows.

Theorem 8.19 Let $G(y)$ be a real valued, positive, continuous function defined on $(-\infty,\infty)$ such that $G(y)=G(-y)$ at all y, and suppose that $G(y)$ is monotonically decreasing for $y\geqslant 0$ and tends to zero as $y\to\infty$: then there will exist a function $f\in L$ whose transform F satisfies

$$|F(y)|>G(y)$$

at all y. □

It is sometimes convenient to form a simplified Fourier inversion theorem which is valid everywhere and can be based on Riemann integrals. Such a theorem may be based either on piecewise monotonic functions (section 5.8) or on piecewise smooth functions (section 5.10). We combine these ideas together as follows.

Theorem 8.20 Let $f(x)$ be a real or complex valued function defined everywhere on $(-\infty,\infty)$, such that,

(i) on each finite interval f is bounded and has only a finite number of discontinuities,
(ii) the following limit of the Riemann integral exists finitely:

$$\lim_{\lambda\to\infty}\int_{-\lambda}^{\lambda}|f(x)|\mathrm{d}x,$$

and either

(iii*a*) f is piecewise smooth on very finite interval, or
(iii*b*) the real and imaginary parts of f are piecewise monotonic on every finite interval: then it follows that at each x the following limits exist and are equal;

$$\tfrac{1}{2}[f(x^+)+f(x^-)]=\lim_{\lambda\to\infty}\int_{-\lambda}^{\lambda}F(y)\exp(2\pi\mathrm{i}xy)\mathrm{d}y \qquad (8.33)$$

where F is defined at all y by,

$$F(y)=\lim_{\lambda\to\infty}\int_{-\lambda}^{\lambda}f(x)\exp(-2\pi\mathrm{i}xy)\mathrm{d}x \qquad (8.34)$$

and the integrals in (8.33) and (8.34) exist as Riemann integrals. □

The condition (i) in this theorem ensures that the proper Riemann integral of f over each finite interval exists, the condition (ii) ensures likewise that F is locally Riemann integrable, whilst the two variants in condition (iii) are based on the Dini and Jordan conditions, respectively. For example, the function $|x|^{1/2}\,\mathrm{e}^{-|x|}$ satisfies the conditions using condition (iii*b*), whilst the function $x^2\cos(1/x)\,\mathrm{e}^{-|x|}$ $(=0$ at $x=0)$ satisfies the conditions using condition (iii*a*).

8.3 Convolution and product theorems in L^p

If $f\leftrightarrow F$ and $g\leftrightarrow G$ and at least one member of each pair is L^p for some $p\in[1,2]$, not necessarily the same p in each case, then it will follow that one or other, or both, of the following will be valid:

$$f*g\leftrightarrow FG \qquad \text{(convolution theorem)} \tag{8.35}$$

$$fg\leftrightarrow F*G \qquad \text{(product theorem).} \tag{8.36}$$

Bearing in mind the close relationship between convolution and cross correlation, this means that at least one of the following will be valid, where ρ_{fg} is defined as in (3.3)

$$\rho_{fg}\leftrightarrow F^*G \tag{8.37}$$

$$f^*g\leftrightarrow \rho_{FG}. \tag{8.38}$$

In certain cases evaluation of $\rho_{fg}(0)$ and use of (8.37) will lead to the Parseval formula,

$$\int f^*g=\int F^*G. \tag{8.39}$$

In this section we describe the various conditions under which (8.35) to (8.39) are valid.

The essence of these theorems is that they place constraints on the behaviour at infinity of each function appearing in a convolution and they place constraints on the local discontinuities of the functions appearing in a product. In all cases awkward behaviour on the part of one function can be compensated by suitable conditions applied to the other function in the product or convolution.

If f and g are both square integrable then each of (8.35) to (8.39) will be valid; this is however only a special case of the following more general result.

Theorem 8.21 Suppose $f\leftrightarrow F$ and $g\leftrightarrow G$, where $f\in L^p$ and $G\in L^p$ for some p satisfying $1\leqslant p\leqslant 2$: then it follows that (i) $fg\in L$ and $FG\in L$, (ii) the convolutions $f*g$ and $F*G$ are continuous everywhere, (iii) at all x and y, respectively,

$$f*g(x)=\int_{-\infty}^{\infty} F(y)G(y)\exp(2\pi ixy)dy \tag{8.40}$$

and

$$F*G(y)=\int_{-\infty}^{\infty} f(x)g(x)\exp(-2\pi ixy)dx, \tag{8.41}$$

(iv) the following Parseval formulae are valid:

$$\left.\begin{aligned}\int fg^* &= \int FG^* \\ \int fg &= \int F_T G\end{aligned}\right\}, \tag{8.42}$$

where F_T is the transpose of F. □

The case $f=g$ may be extracted from the above as an important special case, giving the following.

Theorem 8.22 If, and only if, $f\in L^2$ will it follow that

$$\int |f|^2=\int |F|^2 \qquad \text{(Parseval)}$$

where F is the Fourier transform of f. □

This form of Parseval's equation can be regarded as a limiting case of the following inequality.*

Theorem 8.23 Suppose $f\leftrightarrow F$ and that $f\in L^p$ for some $p\in[1,2]$: then it follows that

$$\|F\|_q \leqslant \|f\|_p \tag{8.43}$$

where $p^{-1}+q^{-1}=1$ (and $q=\infty$ when $p=1$) and the norms are defined as in section 3.1. □

We now give conditions which are sufficient to ensure validity of the convolution formula, but not necessarily the product formula.

Theorem 8.24 Suppose $f\leftrightarrow F$ and $g\leftrightarrow G$ and that $f\in L^p$ for some $p\in[1,2]$; suppose also that either (*a*) $g\in L^r$ for some $r\in[1,2]$, or (*b*) $G\in L^s$ for some s satisfying $p\leqslant s\leqslant 2$: then it follows that

$$f*g\leftrightarrow FG,$$

where $f*g\in L^P$ and $FG\in L^Q$, and with condition (*a*) above P and Q are given

* For a tighter version of the inequality in theorem 8.23, see Beckner (1975).

by

$$\left.\begin{aligned} P^{-1} &= p^{-1}+r^{-1}-1 \\ Q^{-1} &= 2-p^{-1}-r^{-1}, \end{aligned}\right\} \qquad (8.44a)$$

whilst in case (*b*) P and Q are given by

$$\left.\begin{aligned} P^{-1} &= p^{-1}-s^{-1} \\ Q^{-1} &= 1+s^{-1}-p^{-1}, \end{aligned}\right\} \qquad (8.44b)$$

it being understood that $P^{-1}=0$ means $P=\infty$ and $Q^{-1}=0$ means $Q=\infty$. □

We may note that if $f\in L$, then no other conditions are necessary provided $g\leftrightarrow G$ is a Fourier pair in L^p. We note also that when $p=s$ in condition (*b*), then the theorem collapses into theorem 8.21.

A product theorem is obtained on applying the conditions of theorem 8.24 to the opposite function in each pair. Thus, in summary, we find:

Theorem 8.25 Given $f\leftrightarrow F$ and $g\leftrightarrow G$: then

$$fg\leftrightarrow F*G$$

when

(i) $F\in L^p$, $p\in[1,2]$

and either

(ii*a*) $G\in L^r$, $r\in[1,2]$

or

(ii*b*) $g\in L^s$, $p\leqslant s\leqslant 2$,

in which case $F*G\in L^P$ and $fg\in L^Q$, where P and Q are given by (8.44*a*) or (8.44*b*) as appropriate. □

We can test theorems 8.21 and 8.23–8.25 on the Fourier pairs listed in (8.3)–(8.9). Ignoring for the moment a combination of (8.7) with (8.9) we find that all other combinations satisfy the conditions of theorem 8.21 so that the product and convolution formulae (8.40) and (8.41) are valid. The combination of (8.7) with (8.9) satisfies the convolution theorem 8.24 but not the product theorem 8.25: this failure can be traced to the fact that the y-dependent functions in each pair decay too slowly at infinity for the convolution to be defined.

The Fourier pairs in (8.7) and (8.9) provide an example of a case in which $f\leftrightarrow F$, $g\leftrightarrow G$, and $fg\in L$, and yet the transform of fg is not equal to $F*G$ because this convolution is undefined. In such cases it is sometimes possible to define a form of convolution when use is made of a convergence factor, leading to the following formula, at all y:

$$\int_{-\infty}^{\infty} f(x)g(x)\exp(-2\pi i x y)\mathrm{d}x = \lim_{\lambda\to\infty}\int_{-\infty}^{\infty} F(y')G(y-y')K(y'/\lambda)\mathrm{d}y', \quad (8.45)$$

where the convergence function K satisfies conditions (C1) to (C4) of theorem 8.10. The following theorem provides sufficient conditions for the validity of (8.45).

Theorem 8.26 Suppose $f\leftrightarrow F$ and $g\leftrightarrow G$ and that one or other of the conditions of theorem 8.24 is satisfied so that

$$f*g\leftrightarrow FG,$$

and suppose in addition that at least one of the following conditions is satisfied:

(C1) $f\in L^u$ and $g\in L^v$ for some u and v satisfying $u\in[1,2]$, $u^{-1}+v^{-1}=1$ $(\infty^{-1}\equiv 0)$,
(C2) $f(x)=0$ on $(-\infty, b)$ and $g(x)=0$ on (a,∞), where $a<b$:

then it will follow that (8.45) is valid at all y, and that in particular,

$$\int_{-\infty}^{\infty} f(x)g^*(x)\mathrm{d}x = \lim_{\lambda\to\infty}\int_{-\infty}^{\infty} F(y)G^*(y)K(y/\lambda)\mathrm{d}y. \quad \square(8.46)$$

Equation (8.46) represents a modified form of Parseval's equation. Note that condition (C2) leads, not surprisingly, to the value zero on each side of (8.45).

The combination of the Fourier pair (8.7) with (8.9) fails to satisfy the conditions of theorem 8.26 as it stands. However, if the x-axis is partitioned in such a way as to separate the discontinuities in the x-dependent functions of the Fourier pairs, then theorem 8.26 can be applied to the various portions of the x-functions so created. As a result (8.45) is applicable in this case.

An example of a case in which even (8.45) fails is provided by the pairs $f\leftrightarrow F$ and $g\leftrightarrow G$, where

$$f(x)=\begin{cases} x^{-2/3}\,\mathrm{e}^{-|x|} & x>0 \\ 0, & x\leqslant 0 \end{cases} \quad (8.47)$$

$$g(x)=\begin{cases} |x|^{-2/3}\,\mathrm{e}^{-|x|}, & x<0 \\ 0, & x\geqslant 0. \end{cases} \quad (8.48)$$

In this case $f\in L$ and $g\in L$, and the product fg is everywhere zero; however, the right hand side of (8.45) fails to converge for any value of y.

8.4 Uncertainty principle* and bandwidth theorem

If f and F are Fourier transforms it is not possible that the widths of the graphs of $|f(x)|^2$ and $|F(y)|^2$ can both be made arbitrarily small, a fact which underlies the *bandwidth theorem* of signal theory and the *uncertainty principle* of quantum mechanics. If $|f(x)|^2$ and $|F(y)|^2$ are interpreted as weighting functions then the weighted averages, x_0 and y_0, of x and y will be

$$x_0 = \frac{\displaystyle\int_{-\infty}^{+\infty} x|f(x)|^2 dx}{\displaystyle\int_{-\infty}^{+\infty} |f(x)|^2 dx} \tag{8.49}$$

$$y_0 = \frac{\displaystyle\int_{-\infty}^{+\infty} y|F(y)|^2 dy}{\displaystyle\int_{-\infty}^{+\infty} |F(y)|^2 dy}. \tag{8.50}$$

Corresponding measures of the widths of these weight functions are provided by the second moments about the respective means, and it is convenient to define widths Δx and Δy by

$$(\Delta x)^2 = \frac{\displaystyle\int_{-\infty}^{+\infty} (x-x_0)^2|f(x)|^2 dx}{\displaystyle\int_{-\infty}^{+\infty} |f(x)|^2 dx} \tag{8.51}$$

$$(\Delta y)^2 = \frac{\displaystyle\int_{-\infty}^{+\infty} (y-y_0)^2|F(y)|^2 dy}{\displaystyle\int_{-\infty}^{+\infty} |F(y)^2 dy} \tag{8.52}$$

The essence of the bandwidth and uncertainty theorems now lies in the fact that the product $\Delta x \Delta y$ will never be less than $(4\pi)^{-1}$, as in the following.

Theorem 8.27 Suppose $f \in L^2$ and that $f \leftrightarrow F$, so that also $F \in L^2$, and suppose in addition that $xf(x) \in L^2$ and $yF(y) \in L^2$: it follows that (i) x_0, y_0, Δx and Δy can be defined finitely as in (8.49) to (8.52) and that

$$\Delta x \Delta y \geqslant (4\pi)^{-1}, \tag{8.53}$$

(ii) f and F will each be equal a.e. to absolutely continuous functions f_C and F_C whose derivatives, f'_C and F'_C, are each in L^2, (iii) as $x \to \pm\infty$ so $x|f_C(x)|^2$ and $x|F_C(x)|^2$ tend to zero, (iv) the equality in (8.53) will apply if and only if

* For related results based on entropy as a measure of uncertainty see Beckner (1975).

f_C and F_C are of Gaussian form, which means that for arbitrary complex A, and arbitrary real x_0 and y_0, and arbitrary positive a,

$$f_C(x) = Aa\exp[2\pi i x y_0 - \pi a^2(x-x_0)^2]$$
$$F_C(y) = A\exp[-2\pi i x_0(y-y_0) - \pi a^{-2}(y-y_0)^2]. \quad \square$$

The conditions in theorem 8.27 are in fact equivalent to the condition that $f \leftrightarrow F$, where $(1+x)f(x) \in L^2$ and f is equal a.e. to an absolutely continuous function f_C having a derivative $f'_C \in L^2$. In this way the conditions can be placed entirely onto f.

The role of the Gaussian function in providing the minimum uncertainty product is related to the fact that it is not possible for both a function and its transform to decay arbitrarily rapidly at infinity.

Theorem 8.28 Suppose $f(x)$ is a continuous real valued function with $f(x) = f(-x)$ at all x and suppose that $f(x) = O[\exp(-\pi x^2)]$ as $x \to \pm\infty$; let F be the continuous transform of f and suppose also that $F(y) = O[\exp(-\pi y^2)]$ as $y \to \pm\infty$: then it follows that at all x

$$f(x) = F(x) = A\exp(-\pi x^2) \tag{8.54}$$

where A is an arbitrary constant. $\square$

Theorem 8.27 has the drawback of being inapplicable to many commonly met functions such as $r(x)$, given in (8.2), and $x^{-1}\sin x$, because in these cases the function or its transforms has a divergent second moment, (8.51) or (8.52). In such cases a simpler measure of width, based on quantities W_f and W_F, is useful when f and F are both in L. We define

$$W_f = \int_{-\infty}^{\infty} f(x)\mathrm{d}x / f(0)$$
$$W_F = \int_{-\infty}^{\infty} F(y)\mathrm{d}y / F(0)$$

where $f \leftrightarrow F$: it then follows that

$$W_f W_F = 1.$$

Clearly W_f and W_F will be most useful as width parameters when f and F are real and even and when the graphs of f and F are fairly regular in appearance. We can drop the requirement that f and F shall both be in L if convergence functions are used, as follows.

Theorem 8.28 Suppose $f \leftrightarrow F$ and that at least one of these functions is L^p for some $p \in [1, 2]$, and suppose that $f(x)$ and $F(x)$ are continuous at $x = 0$: then it follows that $W_f W_F = 1$, where

$$W_f = \lim_{\lambda\to\infty} \int_{-\infty}^{\infty} f(x)K(x/\lambda)\mathrm{d}x / f(0) \tag{8.55}$$

$$W_F = \lim_{\lambda\to\infty}\int_{-\infty}^{\infty} F(y)K(y/\lambda)\mathrm{d}y/F(0), \tag{8.56}$$

and where the convergence function K satisfies conditions (C1)–(C4) of theorem 8.10; moreover the use of the convergence factor in (8.55) may be replaced by $\lim_{\lambda\to\infty}\int_{-\lambda}^{\lambda}$ whenever $y=0$ is a Dirichlet point of $F(y)$ or by a simple integral when $f\in L$, with analogous replacements in (8.56) when $x=0$ is a Dirichlet point of $f(x)$ or when $F\in L$. □

The functions $r(x)$, given in (8.2), and $x^{-1}\sin x$ are each usefully covered by this theorem.

8.5 The sampling theorem

A function $f(x)$ whose transform $F(y)$ is zero whenever $|y|$ exceeds a cut-off value a (i.e. F is a bounded support) is often referred to by engineers as a *band-limited function*. Band-limited functions are very 'smooth and regular' on account of the lack of 'high frequency' components in the transform. Indeed a band-limited function is equal a.e. to a continuous function which is everywhere infinitely differentiable, and which obeys a Taylor series expansion over every interval. More than this, a band-limited function is so smooth that a knowledge of the values of $f(x)$ at a set of regularly spaced values of x is sufficient to reconstruct $f(x)$ for all x, provided the spacing of the points is not greater than $(1/2a)$. These values of $f(x)$ are called the *sample values* of $f(x)$, the interval $(1/2a)$ is often called the *Nyquist interval*, and the theorem showing how to construct f from its samples is the *sampling theorem*, as follows.

Theorem 8.29 Suppose $F\in L$ and that, for some fixed $a>0$, $F(y)=0$ whenever $|y|>a$: then F is the transform of a continuous function f and, for $n=0, \pm 1, \pm 2, \ldots$, putting $S_n=f(n/2a)$ it will follow that at all x,

$$f(x) = \lim_{N\to\infty}\sum_{n=-N}^{+N} S_n\left[\frac{\sin(2\pi ax-\pi n)}{2\pi ax-\pi n}\right]. \quad \square \tag{8.57}$$

Note that there is no need to consider $F\in L^2$ separately, because this is included in the above theorem due to F having bounded support. Since the sample values S_n determine f completely, they must also determine F completely, and this is achieved as follows.

Theorem 8.30 Suppose that F satisfies the conditions of theorem 8.29 and that f, a and the S_n are as defined there: then,

$$\lim_{N\to\infty}\sum_{n=-N}^{N} S_n\left(1-\frac{|n|}{N+1}\right)\exp(-\pi i n y/a) \tag{8.58}$$

will converge to $F(y)$ for a.a.y in $(-a, a)$, to the Lebesgue value of $F(y)$ at each Lebesgue point of F in $(-a, a)$, and to $F(y)$ at each point of continuity of F in $(-a, a)$; morever

$$\lim_{N\to\infty} \sum_{n=-N}^{N} S_n \exp(-\pi i y n/a) \tag{8.59}$$

will converge to the Dirichlet value of F at each *Dirichlet point* of F in $(-a, a)$, though not necessarily a.e. in $(-a, a)$; if in addition $F \in L^p(-a, a)$ for some $p > 1$, and/or $f \in L^r$ for some $r \in [1, 2]$ then (8.59) will converge to $F(y)$ a.e. on $(-a, a)$. □

If (8.58) or (8.59) is evaluated for values of y outside the range $(-a, a)$ then the function of y so formed is periodic with period $2a$. Further results along these lines including a discussion of 'aliasing', are given in sections 15.5 and 16.4. Other related results appear in section 10.3.

8.6 Hilbert transforms and causal functions

A function $f(x)$ that is zero when $x < 0$ is called a *causal* function because in many physical systems the fact that effect cannot precede cause places this constraint on certain functions describing the system. Causality of a function f endows the transform F with certain properties, one of these being that the imaginary part of F is completely determined by a knowledge of its real part and vice versa; in physical contexts the resulting formulae are often called the Kramers–Kronig relations (Kramers, 1927; Kronig, 1942).

In what follows we will frequently need to consider limits such as the following, for some locally integrable function f,

$$\lim_{\varepsilon\to 0+} \left[\int_{x+\varepsilon}^{\infty} \frac{f(x')}{x-x'} \, dx' + \int_{-\infty}^{x-\varepsilon} \frac{f(x')}{x-x'} \, dx' \right]. \tag{8.60}$$

When this limit exists it is called the *Cauchy principal value* around $x' = x$ of the integral and it is written as

$$P \int_{-\infty}^{\infty} \frac{f(x')}{x-x'} \, dx',$$

where throughout we may replace the limit $-\infty$ and $+\infty$ by a and b, when $a < x < b$. The key feature of the Cauchy principal value is that the limit as $\varepsilon \to 0^+$ applies to both integrals simultaneously in (8.60), rather than one by one, so that there is a chance for the infinities to the left and right of $x' = x$ to cancel each other; this can give a finite result even when $f(x')/(x - x')$ is not integrable around $x' = x$ in the Lebesgue sense. Some conditions governing the existence of the Cauchy principle value are as follows.

Theorem 8.31 Suppose $f \in L^p$ for some $p \in (1, \infty)$: then the following converges for a.a.x to define a function $g \in L^p$,

$$g(x) = P\int_{-\infty}^{\infty} \frac{f(x')}{\pi(x-x')}\,\mathrm{d}x', \tag{8.61}$$

and it follows that, for a.a.x,

$$f(x) = P\int_{-\infty}^{+\infty} \frac{-g(x')}{\pi(x-x')}\,\mathrm{d}x'. \quad \square \tag{8.62}$$

The case $f \in L$ is omitted because although a function g will be definable a.e. by (8.61), g will not necessarily be locally integrable, so that (8.62) becomes inapplicable. Two functions f and g related as in (8.61) and (8.62) are called *Hilbert transforms* of each other.

The relevance of Hilbert transforms to Fourier theory comes from the following.

Theorem 8.32 Suppose $f \leftrightarrow F$ and that at least one of these functions is L^p for some $p \in (1, 2]$: then it follows that the other function is L^q, where $q^{-1} + p^{-1} = 1$, and that

$$f(x)\,\mathrm{sgn}(x) \leftrightarrow G(y)$$

where $G(y)$ is defined a.e. by

$$G(y) = P\int_{-\infty}^{+\infty} \frac{F(y')}{\pi\mathrm{i}(y-y')}\,\mathrm{d}y', \tag{8.63}$$

and where f sgn and G belong to the same integration classes as f and F, respectively. $\square$

The cases $f \in L$ and $F \in L$ require special treatment, as follows.

Theorem 8.33 Suppose $f \in L$ and $f \leftrightarrow F$: then at each y, choosing $a > |y|$, it follows that:

$$\int_{-\infty}^{+\infty} f(x)\,\mathrm{sgn}(x)\,\mathrm{e}^{-2\pi\mathrm{i}xy}\,\mathrm{d}x = \lim_{a\to\infty} P\int_{-a}^{a} \frac{F(y')}{\pi\mathrm{i}(y-y')}\,\mathrm{d}y'. \quad \square$$

Theorem 8.34 Suppose $F \in L$ and $f \leftrightarrow F$: then at a.a.y

$$\lim_{a\to\infty} \int_{-a}^{a} f(x)\,\mathrm{sgn}(x)\left(1 - \frac{|x|}{a}\right)\mathrm{e}^{-2\pi\mathrm{i}xy}\,\mathrm{d}x = P\int_{-\infty}^{+\infty} \frac{F(y')}{\pi\mathrm{i}(y-y')}\,\mathrm{d}y',$$

however the function of y so defined will not necessarily be locally integrable. $\square$

We now turn to a causal function f, in which case the transforms of f and f sgn must be the same. This leads to the real and imaginary parts of the transform F being Hilbert transforms, as follows.

Theorem 8.35 Suppose $f \leftrightarrow F$ and that at least one of these functions is L^r for some $r \in [1, 2]$ and suppose also that $f(x) = 0$ whenever $x < 0$: if we write a.e. $F(y) = X(y) + iY(y)$, where X and Y are real valued, then it will follow that:

(i) whenever $f \in L^p$ for some $p \in (1, 2]$ and/or $F \in L^s$ for some $s \in (1,2]$, then at a.a.y,

$$X(y) = P\int_{-\infty}^{+\infty} \frac{Y(y')}{\pi(y-y')}\,dy' \tag{8.64}$$

and

$$Y(y) = P\int_{-\infty}^{+\infty} \frac{-X(y')}{\pi(y-y')}\,dy', \tag{8.65}$$

(ii) whenever $f \in L$ then X and Y can be chosen to be continuous in which case at all y,

$$X(y) = \lim_{a\to\infty} P\int_{-a}^{a} \frac{Y(y')}{\pi(y-y')}\,dy' \tag{8.66}$$

$$Y(y) = \lim_{a\to\infty} P\int_{-a}^{a} \frac{-X(y')}{\pi(y-y')}\,dy'. \quad \square \tag{8.67}$$

Equations (8.64) to (8.67) are the Kramers–Kronig equations.

It is not easy to test whether a function F is the transform of a causal function, though the following theorem due to Paley and Wiener goes some way towards this.

Theorem 8.36 Suppose $F \in L^2$: then there will exist a function $\theta(y)$ such that $F(y)\exp[i\theta(y)]$ is the transform of a causal function if, and only if,

$$\int_{-\infty}^{+\infty} \frac{\ln|F(y)|}{1+y^2}\,dy$$

exists infinitely. $\square$

The integral is here a Lebesgue integral, so by implication the modulus of the integrand will also be integrable. Since $\ln(0)$ is undefined, it follows that $F(y)$ can have the value zero on at most a null set, if it is to be the transform of a causal function.

9

Fourier theorems for functions outside L^p

9.1 Introduction

We now show how the theory of Fourier transformation can be extended to cover certain pairs of locally integrable functions when neither member of the pair is L^p for any $p \in [1,2]$. The following are examples of such pairs, valid for $a>0$, and $0<\alpha<1$. The gamma function, $\Gamma(\alpha)$, is tabulated for instance in Abramowitz and Stegun (1966).

$$|x|^{-1/2} \leftrightarrow |y|^{-1/2} \tag{9.1}$$

$$|x|^{\alpha-1} \leftrightarrow 2\Gamma(\alpha)\cos(\alpha\pi/2)|2\pi y|^{-\alpha} \tag{9.2}$$

$$|x|^{\alpha-1}\operatorname{sgn} x \leftrightarrow -2i\Gamma(\alpha)\sin(\alpha\pi/2)|2\pi y|^{-\alpha}\operatorname{sgn} y \tag{9.3}$$

$$\cos(ax^2) \leftrightarrow \left(\frac{\pi}{2a}\right)^{1/2}\left[\cos\left(\frac{\pi^2 y^2}{a}\right)+\sin\left(\frac{\pi^2 y^2}{a}\right)\right] \tag{9.4}$$

$$\sin(ax^2) \leftrightarrow \left(\frac{\pi}{2a}\right)^{1/2}\left[\cos\left(\frac{\pi^2 y^2}{a}\right)-\sin\left(\frac{\pi^2 y^2}{a}\right)\right] \tag{9.5}$$

$$\sum_{n=-\infty}^{\infty} n^2\cos(2\pi nx)\exp[-\pi(x-n)^2]$$
$$\leftrightarrow \sum_{n=-\infty}^{\infty} n^2\cos(2\pi ny)\exp[-\pi(y-n)^2] \tag{9.6}$$

$$\sum_{n=-\infty}^{\infty} 2^{|n|}\cos(2\pi nx)\exp[-\pi(x-n)^2]$$
$$\leftrightarrow \sum_{n=-\infty}^{\infty} 2^{|n|}\cos(2\pi ny)\exp[-\pi(y-n)^2] \tag{9.7}$$

The summations in (9.6) and (9.7) converge pointwise to define continuous functions, when $\sum_{n=-\infty}^{\infty}$ is interpreted as $\lim_{N\to\infty}\sum_{n=-N}^{N}$.

The methods used in this chapter form a natural bridge between the classical methods used in previous chapters and the method of generalized functions introduced in chapter 12. We start by considering functions which, although they are not necessarily integrable over $(-\infty, \infty)$, become so when divided by a polynomial of high enough degree; examples (9.1)–(9.6) above are in this category. We then consider functions which are still not integrable when divided by any polynomial, and the Fourier pair (9.7) is of this type.

9.2 Functions in class *K*

The functions which are integrable when divided by a suitable polynomial are conveniently referred to as belonging to class K, this class being more precisely defined as follows.

Definition A locally integrable function f is said to belong to class K if there exists a positive integer N such that $\int_{-\infty}^{\infty} |f(x)|(1+|x|)^{-N}\mathrm{d}x$ exists finitely. □

Equivalent definitions are obtained if the term $(1+|x|)^{-N}$ is replaced by $(1+x^2)^{-N}$ or by $(1+|x|^N)^{-1}$. As examples the functions x^3, $|x|^{-1/2}$, $\ln|x|$ and $\cos x$ are each in class K.

There is a close relationship between the class K and the class S of good functions defined in section 7.1; this arises as follows.

Theorem 9.1 A function f will belong to class K if, and only if, $f\varphi \in L$ for each good function φ. □

When a function f is in class K it is not satisfactory to define a Fourier transform by the convergence at a.a.y of a limit such as

$$\lim_{\lambda\to\infty} \int_{-\lambda}^{\lambda} f(x)\exp(-2\pi \mathrm{i}xy)\mathrm{d}x \tag{9.8}$$

or

$$\lim_{\lambda\to\infty} \int_{-\infty}^{\infty} f(x)\exp(-|x|/\lambda)\exp(-2\pi \mathrm{i}xy)\mathrm{d}x. \tag{9.9}$$

The drawback of (9.8) is that it may diverge at all y, even when a satisfactory transform may be defined by (9.9), as is the case when f is the inverse transform of a Kolmogoroff function, see section 8.2. The drawback of (9.9) is that it may converge at a.a.y to define a locally integrable function that is not a satisfactory transform of f; for example, if $f(x) = 1$ at all x, then (9.9) converges to zero at a.a.y.

There are several equivalent methods of defining Fourier pairs when both the functions are in class K, and we will adopt the following definition based on Parseval's equation rather than a Fourier integral.

Definition If two locally integrable functions f and F are such that

$$\int_{-\infty}^{\infty} f\varphi^* = \int_{-\infty}^{\infty} F\Phi^* \tag{9.10}$$

for every Fourier pair $\varphi \leftrightarrow \Phi$ of good functions (see theorem 7.1) then we say that f and F form a Fourier pair in K and write $f \leftrightarrow F$, F being the Fourier transform of f, and f being the inverse transform of F. □

The definition is unique in the usual sense, in that if $f \leftrightarrow F$ and $f \leftrightarrow G$ in K then $F = G$ a.e. and if $f \leftrightarrow F$ and $g \leftrightarrow F$ in K then $f = g$ a.e. The definition is also consistent with previous definitions for functions in L^p.

The definition based on (9.10) has the drawback of not indicating whether or not a particular $f \in K$ will belong to a pair in K, and also the drawback of not showing directly how to calculate F from f or vice versa. This second point is covered by the following.

Theorem 9.2 Suppose $f \leftrightarrow F$ is a Fourier pair in K, and let Φ be a real valued good function such that $\Phi(0) = 1$ and $\Phi(x) = \Phi(-x)$ at all x: then it follows that

$$\lim_{\lambda \to \infty} \int_{-\infty}^{\infty} f(x)\Phi(x/\lambda) \exp(-2\pi i x y) \mathrm{d}y \tag{9.11}$$

converges to $F(y)$ on the Lebesgue set of F, and so at a.a.y, and also converges to $F(y)$ in senses (ii)–(v) of theorem 8.10; it also follows that

$$\lim_{\lambda \to \infty} \int_{-\infty}^{\infty} F(y)\Phi(y/\lambda) \exp(2\pi i x y) \mathrm{d}y \tag{9.12}$$

converges to $f(x)$ on the Lebesgue set of $f(x)$, and so to $f(x)$ at a.a.x, and also converges to $f(x)$ in senses (ii)–(v) of theorem 8.10 (with F replaced by f). □

A convolution and product theorem for Fourier pairs in K is as follows.

Theorem 9.3 Suppose that $f \leftrightarrow F$ is a Fourier pair in K and that $\varphi \leftrightarrow \Phi$ is any Fourier pair of good functions: it follows that the convolutions $f * \varphi$ and $F * \Phi$ are continuous and everywhere infinitely differentiable, and that at all x and y, respectively

$$f * \varphi(x) = \int_{-\infty}^{+\infty} F(y)\Phi(y)\, \mathrm{e}^{+2\pi i x y}\, \mathrm{d}y \tag{9.13}$$

and

$$F * \Phi(y) = \int_{-\infty}^{+\infty} f(x)\varphi(x)\, \mathrm{e}^{-2\pi i x y}\, \mathrm{d}x. \quad \square \tag{9.14}$$

One way of generating Fourier pairs in K is by the consideration of the limit of a sequence of known Fourier pairs. For example, the function f_a, defined for $a>0$ by

$$f_a(x)=|x|^{-1/2}\exp(-a|x|) \tag{9.15}$$

is in L, and a standard integral (Gradshtein and Ryzhik, 1965), gives its transform F_a as follows, using (8.1),

$$F_a(y)=\left\{\frac{2\pi[a+(a^2+4\pi^2y^2)^{1/2}]}{a^2+4\pi^2y^2}\right\}^{1/2}. \tag{9.16}$$

On taking the limit as $a\to 0^+$ we obtain at all x and y except $x=0$ and $y=0$,

$$\lim_{a\to 0} f_a(x)=|x|^{-1/2}, \qquad \lim_{a\to 0} F_a(y)=|y|^{-1/2},$$

thus suggesting that $|x|^{-1/2}\leftrightarrow|y|^{-1/2}$ in K. Although this result is correct, pointwise a.e. convergence of sequences of Fourier pairs is not sufficient to ensure that the limit functions belong to a Fourier pair; this has already been demonstrated by the pair given in equations (6.11) and (6.12).

One way of ensuring that such a limiting process *does* lead to a valid transform is to replace convergence a.e. by a special kind of dominated convergence, as in the following .

Theorem 9.4 Consider sequences of functions f_n and F_n, $n=1,2,3,\ldots$, such that $f_n\leftrightarrow F_n$ in K for each n, and suppose also that (i) as $n\to\infty$ so f_n and F_n tend pointwise a.e. to functions f and F and that (ii) there exist positive valued functions g and h in K such that for all n

$$|f_n(x)|\leqslant g(x) \quad \text{a.e.,}$$

and

$$|F_n(y)|\leqslant h(x) \quad \text{a.e.:}$$

then it will follow that $f\leftrightarrow F$ in K. □

Throughout this theorem the discrete parameter n may be replaced by a continuous parameter $a>0$, and the limit $n\to\infty$ replaced by $a\to\infty$ or by $a\to 0$. In the example given above one finds that f_a, (9.15), is bounded by $g(x)=|x|^{-1/2}$, whilst for $a>0$ and $y\neq 0$ the function $F_a(y)$, (9.16), is bounded by $h(y)=(2/|y|)^{1/2}$; the conditions of theorem 9.4 are thus satisfied, justifying the pair $|x|^{-1/2}\leftrightarrow|y|^{-1/2}$.

In like manner the pair in (9.2) can be obtained from the limit as $a\to 0^+$ of the following pair (established by a standard integral);

$$|x|^{\alpha-1}\exp(-a|x|)\leftrightarrow\frac{2\Gamma(\alpha)\cos[\alpha\arctan(2\pi y/a)]}{(a^2+4\pi^2y^2)^{\alpha/2}}, \tag{9.17}$$

where $0<\alpha<1$ and the arctangent is defined to have a value within $(-\pi/2,\pi/2)$. The Fourier pair (9.3) is obtained by considering the limit as

$a \to 0^+$ of $|x|^{\alpha-1} e^{-a|x|} \operatorname{sgn} x$. The pairs (9.4) and (9.5) can be derived from the limits as $a \to 0^+$ of the following, for $b > 0$:

$$\exp(-ax^2)\cos bx^2 \leftrightarrow \exp\left(\frac{-a\pi^2 y^2}{A^2}\right)\left[B\cos\frac{b\pi^2 y^2}{A^2} + C\sin\frac{b\pi^2 y^2}{A^2}\right]$$

$$\exp(-ax^2)\sin bx^2 \leftrightarrow \exp\left(\frac{-a\pi^2 y^2}{A^2}\right)\left[C\cos\frac{b\pi^2 y^2}{A^2} - B\sin\frac{b\pi^2 y^2}{A^2}\right]$$

where

$$A = (a^2 + b^2)^{1/2},$$

$$B = \left[\frac{\pi(A+a)}{2A^2}\right]^{1/2}, \qquad C = \left[\frac{\pi(A-a)}{2A^2}\right]^{1/2}.$$

The Fourier pairs (9.6) and (9.7) are built up from the following symmetrical pair of good functions, valid for each $n = 1, 2, 3, 4, \ldots$,

$$\cos(2\pi nx)\{\exp[-\pi(x-n)^2] + \exp[-x+n)^2]\}$$
$$\leftrightarrow \cos(2\pi ny)\{\exp[-\pi(y-n)^2] + \exp[-\pi(y+n)^2]\}. \tag{9.18}$$

Each of the sequences of functions in the pair (9.6) is dominated as $N \to \infty$ by a suitable function in K. For (9.7), however, see section 9.4.

9.3 Convolutions and products in *K*

Given two Fourier pairs $f \leftrightarrow F$ in K and $g \leftrightarrow G$ in K, we now seek sets of conditions which are sufficient to ensure that $f * g \leftrightarrow FG$ in K (convolution formula) or that $fg \leftrightarrow F * G$ in K (product formulae). A rough rule is that constraints on the behaviour at infinity are required on the functions being convoluted together, whilst conditions on local discontinuities are required on the two functions being multiplied together. Also the more stringent is the condition applied to one function, then the less stringent is the condition that is applied to the function with which it is convoluted or multiplied.

Many sets of conditions leading to the convolution formula exist, and we describe some of the most useful. In this description it is useful to introduce a set of functions p_s, for each real s, each function being defined at all real x by,

$$p_s(x) = (1 + |x|)^s. \tag{9.19}$$

The statement $f = O(p_s)$ for some stated value of s means that f is an ordinary function and that there exists an $A > 0$ such that at all x

$$|f(x)| < Ap_s(x). \tag{9.20}$$

The statement that $f = O(p_s)$ at infinity (or as $x \to \pm\infty$) means that for the stated value of s there exists an $a > 0$ and an $A > 0$ such that (9.20) is valid whenever $|x|>a$. The statement $f\in L^p$ at infinity means that for some $a>0$, $f\in L^p(a,\infty)$ and $f\in L^p(-\infty,-a)$.

We start with a convolution theorem in which the conditions are confined to the x or y domain.

Theorem 9.5 Suppose $f\leftrightarrow F$ in K and $g\leftrightarrow G$ in K, and that at least one of the following conditions is satisfied:

(C1) $fp_s\in L$ for all $s>0$,
(C2) G is infinitely differentiable and $G^{(k)}\in L^2$ for each $k=1,2,3,\ldots,$
(C3) there exists an $s\geqslant 0$ or an $s<-1$ such that $fp_s\in L$ and $g=O(p_s)$ at infinity:

then it follows that $f*g\leftrightarrow FG$ in K. □

For example, if we put $f(x)=|x|^{-1/2}\,\mathrm{e}^{-|x|}$ then (C1) is satisfied, and the other pair $g\leftrightarrow G$ can be any pair chosen from (9.1) to (9.6). Likewise if $G(y)=(1+\mathrm{i}y)^{-1}$ then (C2) is satisfied and there are no constraints on the other pair in K. A third example is provided by choosing $f(x)=\cos ax^2$ and $g(x)=|x|^{-1/2}(1+|x|)^{-1}$, in which case condition (C3) is satisfied, with $s=-3/2$.

The following is a more general theorem in which conditions are placed on the x- and y-dependent functions. The theorem is not easy to apply directly but leads to specific cases which we will list afterwards.

Theorem 9.6 Suppose $f\leftrightarrow F$ in K and $g\leftrightarrow G$ in K and that conditions (CX) and (CY) below are both satisfied,

(CX) $f_{\mathrm{T}}(|g|*|\varphi|)\in L$ for every good φ (where f_{T} is the transpose of f),
(CY) $(\varphi F)*G_{\mathrm{T}}$ is continuous at the origin for every good φ (where G_{T} is the transpose of G):

then it follows that $f*g\leftrightarrow FG$. □

One can see that (CX) places conditions on the behaviour of f and g at infinity, local discontinuities being irrelevant, whilst (CY) places constraints on the way infinite discontinuities in F and G superpose. The following theorems provide more explicit tests replacing (CX) and (CY). We give examples afterwards.

Theorem 9.7 In order that $f\in K$ and $g\in K$ shall satisfy condition (CX) above it is sufficient that at least one of the following is satisfied:

(CX1) there exists an $s\in(-\infty,\infty)$ such that $fp_s\in L$ and that $g=O(p_s)$ at infinity,

(CX2) $f \in L^p$ at infinity for some $p \in [1, \infty]$ and $g \in L^q$ at infinity for some $q \in [1, \infty]$ where $1 \leqslant p^{-1} + q^{-1} \leqslant 2$ $(\infty^{-1} \equiv 0)$,

(CX3) there exists a real b such that *either* $f(x) = g(x) = 0$ on $(-\infty, b)$ *or* $f(x) = g(x) = 0$ on (b, ∞). □

Theorem 9.8 In order that $F \in K$ and $G \in K$ shall satisfy condition (CY) of theorem 9.6 it is sufficient that at least one of the following is satisfied:

(CY1) there exists an $s > 0$ such that $|F| p_{-s}$ is bounded on $(-\infty, \infty)$,

(CY2) there exist s, t, p, q such that $Fp_{-s} \in L^p$ and $Gp_{-t} \in L^q$, where s, t, p and q satisfy $s \geqslant 0, t \geqslant 0, p \in [1, \infty], q \in L[1, \infty], 0 \leqslant p^{-1} + q^{-1} \leqslant 1$ and as usual $\infty^{-1} \equiv 0$.

(CY3) $G(y) = 0$ on $(-\infty, b)$ and $F(y) = 0$ on (a, ∞) for some finite $a < b$. □

As an example drawn from (9.2) we can establish the following, with $f(x) = |x|^{-2/3}$ and $g(x) = |x|^{-3/4}$,

$$|x|^{-2/3} * |x|^{-3/4} \leftrightarrow \text{constant } |y|^{-1/3} |y|^{-1/4}.$$

This follows because condition (CX1) is satisfied (with $s = -\frac{3}{4}$) and condition (CY2) is satisfied (with $s = 2$, $t = 2$, $p = 2$, $q = 3$).

Conditions for the validity of a product formula, $fg \leftrightarrow F * G$, are obtained from theorems 9.5–9.8 by applying the various conditions on f and g to F and G and vice versa.

9.4 Functions outside *K*

The approach adopted in sections 9.2 and 9.3 may be adapted to deal with transforms of functions not in K if the good functions are replaced throughout by certain restricted classes of good functions. For example, if throughout we use only those good functions, φ, for which $\varphi(x) = O(e^{-\lambda|x|})$ for all $\lambda > 0$ then we may replace the class K by the class of functions, f, such that $f(x)\,e^{-\lambda|x|}$ is in L for some $\lambda > 0$. For instance the good functions $\exp(-x^2)$ and $x\exp(-x^2)$ are in this new category, and the functions f and F in (9.7) form an integrable product with these good functions although f and F are not in K. The pair (9.7) is justified by a modified version of theorem 9.4. Indeed all the theorems in sections 9.2 and 9.3 can be adapted to this approach, though we will not follow it through. One is restricted in how fast one makes the good functions decay at infinity on account of theorem 8.28, which says that a Gaussian decay is the best possible if symmetry between the x and y domains is retained. However, if symmetry is abandoned we can choose the good functions $\varphi(x)$ to be of bounded support: in this case $f\varphi \in L$ for each function f that is locally integrable. A very common approach is based on the class D, the good functions of bounded support, and we describe these results in further detail; they provide a useful background to the generalized functions called distributions.

We extend the definition of a Fourier pair as follows.

Definition Suppose f and F are locally integrable functions such that for every $\varphi \in D$,

$$\int f\varphi^* = \int F\Phi^* \tag{9.21}$$

where $\varphi \leftrightarrow \Phi$; we then define f and F as a *regular pair in D'* and write $f \leftrightarrow F$: alternatively if (9.21) holds for every $\Phi \in D$, and $\varphi \leftrightarrow \Phi$, then we define f and F as a regular pair in Z' and write $f \leftrightarrow F$. □

This definition is consistent with earlier definitions, in that if $f \leftrightarrow F$ in K, then it follows that $f \leftrightarrow F$ is a regular pair in D' and in Z'. Moreover if $f \leftrightarrow F$ is a regular pair in either D' or Z', and if $f \in K$ and $F \in K$, then it follows that $f \leftrightarrow F$ in K. The regular pairs in D' constitute a linear transformation, (6.3), as also do those in Z'; however, if $f \leftrightarrow F$ is a regular pair in D' and $g \leftrightarrow G$ is a regular pair in Z', it will not necessarily follow that $(f+g) \leftrightarrow (F+G)$ form a regular Fourier pair in D' or in Z'. This follows from the unsymmetrical nature of the definition between the x and y domains. The theorems that follow will be phrased so as to apply to regular pairs in D', but it is clear that by interchanging the roles of D and Z we obtain an analogous set of theorems for regular pairs in Z'.

The basic Fourier inversion theorem now becomes:

Theorem 9.9 Suppose $f \leftrightarrow F$ is a regular pair in D', and let $\varphi \leftrightarrow \Phi$ be a pair with $\varphi \in D$ and $\Phi \in Z$ such that φ and Φ are real valued and $\varphi(0) = \Phi(0) = 1$ and $\varphi(x) = \varphi(-x)$ at all x: then

$$\lim_{\lambda \to \infty} \int_{-\infty}^{+\infty} f(x)\varphi(x/\lambda) \exp(-2\pi i x y) \mathrm{d}x \tag{9.22}$$

and

$$\lim_{\lambda \to \infty} \int_{-\infty}^{+\infty} F(y)\Phi(y/\lambda) \exp(2\pi i x y) \mathrm{d}y \tag{9.23}$$

will converge to $F(y)$ and $f(x)$, respectively, in the ways described in theorem 9.2 under (9.11) and (9.12). □

The basic convolution and product theorem is now as follows.

Theorem 9.10 Suppose $f \leftrightarrow$F is a regular pair in D', and let $\varphi \leftrightarrow \Phi$ be a pair with $\varphi \in D$ and $\Phi \in Z$: then the convolutions $f * \varphi$ and $F * \Phi$ are each continuous functions, and at all x and y, respectively

$$f*\varphi(x)=\int_{-\infty}^{+\infty} F(y)\Phi(y)\exp(2\pi ixy)\mathrm{d}y$$

and

$$F*\Phi(y)=\int_{-\infty}^{+\infty} f(x)\varphi(x)\exp(-2\pi ixy)\mathrm{d}x. \quad \square$$

One way of constructing regular Fourier pairs in D' is by means of sequences of established pairs, as follows.

Theorem 9.11 Consider sequences of functions f_n and F_n, $n=1,2,3,\ldots$, such that $f_n\leftrightarrow F_n$ is a regular pair in D' for each n, and suppose that (i) as $n\to\infty$ so f_n and F_n tend pointwise a.e. to functions f and F, and that (ii) there exists a positive function $g\in L_{\mathrm{LOC}}$ and a positive function $h\in K$ such that for all n

$$|f_n(x)|\leqslant g(x) \quad \text{a.a.}x$$

$$|F_n(y)|\leqslant h(y) \quad \text{a.a.}y:$$

then it will follow that $f\leftrightarrow F$ is a regular pair in D'. $\square$

The discrete parameter may, as in theorem 9.4, be replaced by a continuous parameter. As an example of the use of this theorem consider the pairs $f_n\leftrightarrow F_n$, $n=1,2,3,\ldots$, defined as follows:

$$f_n(x)=\sum_{m=1}^{n} A_m\cos(2\pi A_m x)\{\varphi(x-m)+\varphi(x+m)\} \tag{9.24}$$

$$F_n(x)=\sum_{m=1}^{n} A_m\cos(2\pi my)\{\Phi(y-A_m)+\Phi(y+A_m)\}, \tag{9.25}$$

where A_m is the integer equal to 2 raised to the power 2^m, and where $\varphi\in D$ is the function

$$\varphi(x)=\begin{cases}\exp[-(1-4x^2)^{-1}] & |x|\leqslant\frac{1}{2}\\ 0 & |x|>\frac{1}{2}\end{cases}$$

and Φ is the transform of φ. The functions f_n and F_n are in fact all good functions and are readily established as Fourier pairs using the shift theorem, (6.4). As $n\to\infty$ so f_n and F_n tend everywhere to continuous limit functions f and F, and moreover the sequences are bounded by $g(x)=\exp(\exp|x|)$ and by $h(y)=2|y|$, respectively. Since $g\in L_{\mathrm{LOC}}$ and $h\in K$, the conditions of theorem 9.11 are met, and consequently the pair

$$\sum_{m=-\infty}^{\infty} A_m\cos(2\pi A_m x)\varphi(x-m)\leftrightarrow \sum_{m=-\infty}^{\infty} A_m\cos(2\pi my)\Phi(y-A_m) \tag{9.26}$$

is a regular Fourier pair in D'. In fact the integers A_m could be replaced by any other set increasing arbitrarily fast with increase in m, and likewise φ

could be chosen as any function in D. It is clear that in this example f is not in K, so f and F do not form a Fourier pair in K.

A convolution theorem can be based on functions of bounded support as follows.

Theorem 9.12 Suppose $f \leftrightarrow F$ is a regular pair in D', and suppose $g \in L$ is of bounded support and that $g \leftrightarrow G$: then it will follow that $f * g \leftrightarrow FG$ is a regular pair in D'. □

For example, if the left hand function in (9.26) is convoluted with a rectangle function (8.2), then the regular transform in D' will be equal to the product of the right hand side of (9.26) with the function $(\pi y)^{-1} \sin(2\pi y)$.

A product theorem can be constructed as follows.

Theorem 9.13 Suppose $f \leftrightarrow F$ is a regular pair in D' with $F \in K$, and suppose a function G is such that $G(y)(1+|y|)^N \in L$ for all $N \geqslant 0$, and that $g \leftrightarrow G$: then it will follow that $fg \leftrightarrow F * G$ is a regular pair in D'.□

For example, if $f \leftrightarrow F$ are as in (9.26) and $G(y) = \exp(-|y|)$ then the conditions of theorem 9.13 will be satisfied, and $fg \leftrightarrow F * G$.

A feature of the Fourier pairs in (9.6), (9.7) and (9.26) is that the functions oscillate as $x \to \pm\infty$. This is related to the following.

Theorem 9.14 Suppose that $f \leftrightarrow F$ is a Fourier pair in K or a regular pair in D' or Z', and suppose that $f(x) \geqslant 0$ at a.a.x: then it follows that for all $s > 1$,

$$\int_{-\infty}^{\infty} \frac{|f(x)|}{(1+|x|)^s} \, \mathrm{d}x < \infty. \quad \Box \tag{9.27}$$

This result implies that if $f \leftrightarrow F$ in one of the above senses and if the integral in (9.27) diverges for some $s > 1$ then f cannot be everywhere real and positive.

10

Miscellaneous theorems

10.1 Differentiation and integration

The rule according to which $f'(x) \leftrightarrow 2\pi \mathrm{i} yF(y)$ when $f(x) \leftrightarrow F(y)$ applies to functions other than the good functions. The rule can apply even when the derivative f' is undefined at certain points, though also the rule can fail when f' is defined everywhere. The crucial concept, within the realm of integrable functions, is that of absolute continuity rather than simply differentiability.

One can place conditions either on f or on F, thus leading to two types of differentiation theorem, as follows.

Theorem 10.1 Let $f \leftrightarrow F$ be a Fourier pair in D' or in Z' and suppose that (i) f is an indefinite integral of its derivative f' on every finite interval and that (ii) at least one out of f' and F is in class K: then it will follow that

$$f'(x) \leftrightarrow 2\pi \mathrm{i} yF(y)$$

is a Fourier pair of the same type (D' or Z') as the pair $f \leftrightarrow F$. □

For example, on differentiating the left hand member of

$$\exp(-|x|) \leftrightarrow 2(1+4\pi^2y^2)^{-1}$$

we obtain

$$-\exp(-|x|)\operatorname{sgn} x \leftrightarrow \frac{4\pi \mathrm{i} y}{1+4\pi^2y^2}.$$

On placing the conditions on F we get

Theorem 10.2 Let $f \leftrightarrow F$ be a Fourier pair in D' or Z' and suppose that $2\pi \mathrm{i} yF(y)$ belongs to a Fourier pair in D' or Z' (irrespective of the class of $f \leftrightarrow F$): then it follows (i) that f is equal a.e. to an absolutely continuous function f_C, and (ii) that

$$f_C'(x) \leftrightarrow 2\pi \mathrm{i} yF(y)$$

is a Fourier pair of the same type (D' or Z') as $f \leftrightarrow F$, and (iii) that at a.a. x

$$f(x) = f_C(0) + \int_0^x f'_C(u)\mathrm{d}u. \quad \square$$

For example, choosing $F(y) = (1+y^2)^{-1}$, we find that $F \in L(-\infty, \infty)$, and that $yF(y) \in L^2(-\infty, \infty)$, so that the conditions of theorem 10.2 are satisfied since both $F(y)$ and $(2\pi \mathrm{i} y)F(y)$ must belong to Fourier pairs in L^p (and thus in D' and Z').

Clearly by repeated application of the above theorems one builds up conditions under which for some given n,

$$f^{(n)}(x) \leftrightarrow (2\pi \mathrm{i} y)^n F(y). \tag{10.1}$$

Likewise by reversing the roles of f and F one obtains conditions under which

$$(-2\pi \mathrm{i} x)^n f(x) \leftrightarrow F^{(n)}(y). \tag{10.2}$$

Thus, for example, if $(1+y)^n F(y)$ is either $L(-\infty, \infty)$ or $L^2(-\infty, \infty)$ then (10.1) will be valid, f will be $(n-1)$ times differentiable everywhere and $f^{(n-1)}$ will be absolutely continuous. Conversely, if f is $(n-1)$ times differentiable everywhere, and $f^{(n-1)}$ is absolutely continuous, and if each of $f, f', f'', \ldots, f^{(n)}$ is either $L(-\infty, \infty)$ or $L^2(-\infty, \infty)$ then also (10.1) will be true.

In stating integration theorems we start, for simplicity, with Fourier pairs in L^p and then generalize the result.

Theorem 10.3 Suppose $f \leftrightarrow F$ and that at least one of the functions is $L^p(-\infty, \infty)$ for some $p \in [1, 2]$: then it follows that for each real a and each real x,

$$\int_a^x f(u)\mathrm{d}u = \lim_{\lambda \to \infty} \int_{-\lambda}^{\lambda} F(y) \left[\frac{\exp(2\pi \mathrm{i} xy) - \exp(2\pi \mathrm{i} ay)}{2\pi \mathrm{i} y} \right] \mathrm{d}y;$$

moreover if $f \in L^p$, where $p \in (1, 2]$ or if $F(y)/(1+y)$ is $L(-\infty, \infty)$ then $\lim_{\lambda \to \infty} \int_{-\lambda}^{\lambda}$ may be replaced by $\int_{-\infty}^{\infty}$. $\square$

The content of the square bracket in this equation tends to a finite limit as $y \to 0$ so it is not necessary that $F(y)$ should tend to zero as $y \to 0$.

The use of convergence factors allows Fourier pairs which are not necessarily in L^p to be dealt with.

Theorem 10.4 Suppose $f \leftrightarrow F$ is a regular Fourier pair in D' or in Z': then at each real a and each real x

$$\int_a^x f(u)\mathrm{d}u = \lim_{\lambda \to \infty} \int_{-\infty}^{\infty} F(y)\Phi(y/\lambda) \left[\frac{\exp(2\pi \mathrm{i} xy) - \exp(2\pi \mathrm{i} ay)}{2\pi \mathrm{i} y} \right] \mathrm{d}y,$$

where Φ is a function satisfying (i) $\Phi \in Z$ when $f \leftrightarrow F$ in D', $\Phi \in D$ when $f \leftrightarrow F$

in Z', or $\Phi \in S$ when $f \leftrightarrow F$ in K, and (ii) Φ is real, $\Phi(0)=1$, and $\Phi(-y)=\Phi(y)$ at all y. □

10.2 The Gibbs phenomenon

The Gibbs phenomenon consists in an overshoot which can occur in the Fourier synthesis of a function containing a step discontinuity. An example will introduce it. Consider the function f defined by

$$f(x)=\begin{cases} 1, & -1<x<1 \\ \frac{1}{2}, & x=\pm 1 \\ 0, & |x|>1. \end{cases}$$

This function has the Fourier transform $F(y)=(\pi y)^{-1}\sin 2\pi y$. If we define f_λ by

$$f_\lambda(x)=\int_{-\lambda}^{\lambda} F(y)\exp(2\pi i x y)\mathrm{d}y \tag{10.3}$$

then as $\lambda\to\infty$ so $f_\lambda(x)$ will tend pointwise to $f(x)$ at *all* x (see theorems 8.12 and 8.20). Convergence occurs even at $x=\pm 1$ because we have defined $f(x)$ to have a value 'half way up the step' at these points. However, let us examine the behaviour of $f_\lambda(x)$, as $\lambda\to\infty$, in a region just to one side of one of the discontinuities, choosing for example the interval $x\in(0, 1)$. At fixed λ the error, $f_\lambda(x)-f(x)$, oscillates above and below the value zero as $x\to 1$, reaching a maximum positive value at some point say $x=x_\lambda$: we can call $f_\lambda(x_\lambda)-f(x_\lambda)$ the overshoot. As $\lambda\to\infty$ so the period of the oscillations tends to zero, and so also $x_\lambda\to 1$; however, the value of the overshoot, $f_\lambda(x_\lambda)-f(x_\lambda)$, does *not* tend to zero but instead tends to a finite limit. The existence of this finite, non-zero, limiting value for the overshoot constitutes the Gibbs phenomenon.

We can describe the phenomenon more precisely as follows.

Theorem 10.5A Let $f\leftrightarrow F$ be a Fourier pair of locally integrable functions in any of the senses so far defined and suppose that $f(x)/(1+x)$ is $L(-\infty,\infty)$; suppose also that (i) f is real valued and of bounded variation on some interval $[a, c]$ containing the point $x=b$, $a<b<c$, and (ii) f is continuous on (a, b) and on (b, c), and (iii) $f(b^+)-f(b^-)>0$: then it will follow that with f_λ defined by (10.3),

$$\lim_{\lambda\to\infty}\max_{b<x<c}[f_\lambda(x)-f(x)]$$

$$=\lim_{\lambda\to\infty}\max_{a<x<b}[f(x)-f_\lambda(x)]=A[f(b^+)-f(b^-)], \tag{10.4}$$

where

$$A = \int_0^\pi \frac{\sin u}{\pi u}\, du - \tfrac{1}{2} \approx 0.08949\ldots; \tag{10.5}$$

moreover if the maximum value of $[f_\lambda(x) - f(x)]$ on (b, c) occurs at the point $x = b + \delta_\lambda$, then

$$\lim_{\lambda \to \infty} 2\lambda\delta_\lambda = 1. \quad \square \tag{10.6}$$

The Gibbs phenomenon occurs in an almost identical fashion in the Fourier synthesis of periodic functions using Fourier series. Although Fourier series will only be fully treated in chapters 15 and 16 it is nevertheless convenient to state the relevant theorem now.

Theorem 10.5B Suppose f is of period X, is $L(0, X)$, and has Fourier coefficients $\{F_n\}$, (15.2), and suppose also that on some interval $[a, c]$ f satisfies the conditions (i), (ii) and (iii) given in theorem 10.5A: then if $s_N(x)$ is the partial sum defined by (15.4) it will follow that

$$\lim_{N \to \infty} \max_{b<x<c} [s_N(x) - f(x)]$$

$$= \lim_{N \to \infty} \max_{a<x<b} [f(x) - s_N(x)] = A[f(b^+) - f(b^-)], \tag{10.7}$$

where A has the value given in (10.5); moreover if the maximum value of $[s_N(x) - f(x)]$ on (b, c) occurs at $x = b + \delta_\lambda$, then

$$\lim_{N \to \infty} (2N\delta_\lambda / X) = 1. \quad \square$$

The use of a convergence function as in theorems 8.10, 8.11, 9.2, 9.9 and 15.8 does *not* necessarily eliminate the Gibbs phenomenon, but it does remove the need for the condition (i) (bounded variation) in theorems 10.5A, B and it also eliminates the condition $f(x)/(1+x) \in L$ in theorem 10.5A. However, if the convergence function K in any of these theorems is the transform of a function k that is real and nowhere negative, then the Gibbs phenomenon *is* eliminated: this is the case for instance with the Fejér function, $(1-|x|)$ $[=0, |x|>1]$, and with the exponential function $K(x) = \exp(-|x|)$. If the function k does have negative values then theorem 10.5A can be modified to apply to any Fourier pair to which the convergence function is applicable, as follows.

Theorem 10.6A Suppose the Fourier pair $f \leftrightarrow F$ and the convergence function K jointly satisfy the conditions of one or more of theorems 8.10, 8.11, 9.2 and 9.9 and that the inverse transform, k, of K has negative values on a set of positive measure; suppose also that (i) f is real valued on an interval $[a, c]$ containing the point $x=b$, $a<b<c$, and that conditions (ii) and (iii) of theorem 10.5A are satisfied: then (10.4) will be valid when f_λ is defined at all x by

$$f_\lambda(x)=\int_{-\infty}^{\infty} F(y)K(y/\lambda)\exp(2\pi ixy)\mathrm{d}y$$

and when the constant A has the value

$$A=\max_{0<x<\infty}\left[\int_0^x k(u)\mathrm{d}u-\tfrac{1}{2}\right]. \quad \square \tag{10.8}$$

Theorem 10.5B for periodic functions may be modifed in an analogous manner.

Theorem 10.6B Suppose a function f, of period X, and a convergence function K jointly satisfy the conditions of theorem 15.8, and that the inverse transform, k, of K has negative values on a set of positive measure; suppose also that f satisfies conditions (i), (ii) and (iii) of theorem 10.6A: then (10.4) will be valid when f_λ is redefined as

$$f_\lambda(x)=\sum_{n=-\infty}^{\infty} F_n K(n/\lambda)\exp(2\pi inx/X)$$

and when the constant A has the value in (10.8). $\square$

The use of the Cesàro or Abel method of summing a Fourier series, as in theorem 15.7, eliminates the Gibbs phenomenon.

10.3 Complex Fourier transforms

The complex Fourier transform was encountered in section 7.3 as an aid in characterizing those good functions whose transform (or inverse transform) was zero outside of some finite interval. The same approach can be used with functions other than good functions, and we describe now a cluster of such theorems.

Given a complex valued function $f(x)$ that is defined a.e. on the real line and is locally integrable, we define its complex Fourier transform as the function $F_c(z)$, of the complex variable z, whose value is

$$F_c(z)=\int_{-\infty}^{\infty} f(x)\exp(-2\pi ixz)\mathrm{d}x \tag{10.9}$$

for any value of z at which the integral exists: at other values of z the function F_c is left undefined.

It is convenient to write $z=y+i\alpha$, where y and α are real, and it turns out that $F_c(z)$ will be defined at a point z in the complex plane if, and only if, $f(x)\exp(2\pi\alpha x)$ is $L(-\infty, \infty)$, irrespective of the value of y. Thus the function $f(x)=e^{-2\pi|x|}$ has a complex Fourier transform defined for $-1<\alpha<1$, whilst $f(x)=(1+x^2)^{-1}$ has a complex Fourier transform defined only for $\alpha=0$. Quite often we shall consider $F_c(y+i\alpha)$ as a function of y at fixed α, and it is then useful to introduce the function $F_\alpha(y)$ such that $F_\alpha(y)=F_c(y+i\alpha)$. If also we define f_α by

$$f_\alpha(x)=f(x)\exp(2\pi\alpha x),$$

then (10.9) becomes equivalent to the ordinary transform,

$$F_\alpha(y)=\int_{-\infty}^{+\infty} f_\alpha(x)\exp(-2\pi ixy)dx$$

whenever $f_\alpha \in L(-\infty,\infty)$.

A Fourier inversion theorem for retrieving $f(x)$ from $F_c(z)$ is not essentially different from any of those applicable to the ordinary transform of a function $f\in L$. Thus if x is a Dirichlet point of $f_\alpha(x)$, then

$$\lim_{\lambda\to\infty}\int_{-\lambda}^{\lambda} F_\alpha(y)\exp(2\pi ixy)dy \tag{10.10}$$

will converge to the Dirichlet value of $f_\alpha(x)$ at that point, for any value of α at which F_α is defined. If $f_D(x)$ is the Dirichlet value of f at such a point, (10.10) can be rewritten

$$f_D(x)=\lim_{\lambda\to\infty}\int_{-\lambda}^{\lambda} F_c(y+i\alpha)\exp[2\pi ix(y+i\alpha)]dy \tag{10.11}$$

$$=\lim_{\lambda\to\infty}\int_{-\lambda+i\alpha}^{\lambda+i\alpha} F_c(z)\exp(2\pi ixz)dz, \tag{10.12}$$

where (10.12) is defined to mean the same as (10.11), and where α can be chosen to have any value for which F_α is defined. By introducing a convergence function $K(y)$ into the integrals of (10.10)–(10.12) one can achieve convergence at a.a.x as in theorems 8.10, 8.11, 9.2 and 9.9.

When $f(x)$ is of bounded support, so that for some $b>0$, $f(x)=0$ whenever $|x|>b$, then the complex Fourier transform $F_c(z)$ has special properties. The ordinary transform $F(y)$ is then 'band limited', and, as theorem 8.29 describes, F can be reconstructed from a knowledge of suitably sampled values. The following theorem describes the resulting properties of F_c.

Theorem 10.7 Suppose $f\in L(-\infty,\infty)$ and that for some $b>0$, $f(x)=0$ whenever $|x|>b$: then f has a complex Fourier transform F_c with the following properties: (i) $F_c(z)$ is an entire function, (ii) a constant $C>0$ exists such that at all y and α

$$|F_c(y+i\alpha)|<C\exp(2\pi b|\alpha|), \tag{10.13}$$

and (iii), if $F(y)$ is the ordinary continuous Fourier transform of $f(x)$, then

$$\lim_{\lambda\to\infty}\int_{-\lambda}^{\lambda}\frac{F(u)\exp[-2\pi i b(z-u)]}{2\pi i(z-u)}\,du=\begin{cases}-F_c(z), & \operatorname{Im} z>0\\ 0, & \operatorname{Im} z<0\end{cases} \tag{10.14}$$

and

$$\lim_{\lambda\to\infty}\int_{-\lambda}^{\lambda}\frac{F(u)\exp[+2\pi i b(z-u)]}{2\pi i(z-u)}\,du=\begin{cases}0, & \operatorname{Im} z>0\\ F_c(z), & \operatorname{Im} z<0\end{cases}, \tag{10.15}$$

and (iv) as $\alpha\to 0$, from above or below, so $F_\alpha(y)$ tends uniformly to $F(y)$ on $(-\infty,\infty)$. □

One can embellish the above result by adding extra conditions on f, as follows.

Theorem 10.8 Suppose that f is $L^p(-\infty,\infty)$ for some $p\in(1,2]$ and that $f(x)=0$ when $|x|>b$, for some fixed $b>0$: then the conditions of theorem 10.7 are fulfilled, and in addition to (i)–(iv) above, it will follow that (v) a different constant C exists such that at all y and α

$$\int_{-\infty}^{+\infty}|F_c(y+i\alpha)|^q dy<C\exp(2\pi q b|\alpha|), \tag{10.16}$$

where $p^{-1}+q^{-1}=1$, (vi) as $\alpha\to 0^\pm$ so

$$\text{l.i.m.}(q)F_\alpha(y)=F(y),$$

and (vii) in (10.14) and (10.15) one may replace $\lim_{\lambda\to\infty}\int_{-\lambda}^{\lambda}$ by $\int_{-\infty}^{\infty}$. □

As a converse to the above two theorems we may seek a condition on $F(y)$ which will ensure that it is the transform of a function $f(x)$ of bounded support. The following provides sufficient, but not necessary conditions.

Theorem 10.9 Let $F_c(z)\equiv F_\alpha(y)$ be an entire function such that for some p satisfying $1<p\leqslant 2$ the following holds at all y and α,

$$\int_{-\infty}^{\infty}|F_\alpha(y)|^p dy<C\exp(2\pi p b|\alpha|) \tag{10.17}$$

where C and b are positive constants: then it follows that, (i) $F_c(z)$ is the complex Fourier transform of a function $f\in L^q$ such that $f(x)=0$ whenever $|x|>b$ (where $p^{-1}+q^{-1}=1$), (ii) as $\alpha\to 0^\pm$ so

$$\text{l.i.m.}(p)F_\alpha(y)=F(y)$$

where F is the ordinary Fourier transform of f, and (iii) (10.14) and (10.15) will be valid with $\lim_{\lambda\to\infty}\int_{-\lambda}^{\lambda}$ replaced by $\int_{-\infty}^{\infty}$. □

Note that the conditions in theorems 10.8 and 10.9 become equivalent when $p=2$, thus:

Theorem 10.10 An entire function $F(z) \equiv F_\alpha(y)$ will be the complex Fourier transform of a function $f \in L^2$, such that $f(x) = 0$ whenever $|x| > b$, if and only if there exists a $C > 0$ such that (10.17) is valid at all α when $p = 2$. □

The following provides an alternative way of placing conditions on F_c.

Theorem 10.11 Let $F_c(z)$ be an entire function such that for some positive integer k, and some positive constants C and b,

$$|z|^k F_c(z) < C \exp(2\pi b |\mathrm{Im}\, z|) \tag{10.18}$$

at all z: then the conditions of theorem 10.9 are satisfied for all $p \in (1, 2]$; moreover when $k = 2$ then f will be absolutely continuous, and when $k > 2$ then f will be $(k-2)$ times differentiable with the derivative $f^{(k-2)}$ absolutely continuous. □

One can see how the specification of a good function in class Z, theorem 7.7, follows as a natural extension of the above theorem.

Theorem 10.11 fails when $k = 0$, since for example the entire function $F_c(z) = 1$ (all z) is not the complex transform of any ordinary function. However, within the realm of generalized functions, an extension of theorem 10.11 will be useful in specifying certain generalized functions of bounded support (see theorem 14.3).

We conclude with some results relating to the complex Fourier transform of a causal function. A function $f(x)$ is defined as being causal if $f(x) = 0$ almost everywhere on $(-\infty, 0)$. The complex Fourier transform $F_c(z)$ is not necessarily an entire function in this case, but is analytic on the half plane $\mathrm{Im}\, z < 0$. By this we mean that $F_c(y + \mathrm{i}\alpha)$ is differentiable at every point for which $\alpha < 0$; this means in turn that at every such point F_c is also continuous and infinitely differentiable. It is important that we exclude the boundary value $\alpha = 0$ in the above remarks. Some authors, including Titchmarsh (1962), call such functions analytic and regular on the half plane, the word analytic by itself then allowing certain types of singularity at a finite number of points.

The following two theorems place conditions on f and F_c, respectively.

Theorem 10.12 Suppose $f \in L^p(-\infty, \infty)$ for some p satisfying $1 \leqslant p \leqslant 2$ and that $f(x) = 0$ when $x < 0$: then f has a complex Fourier transform $F_c(z)$ that is analytic on the half plane $\mathrm{Im}\, z < 0$ and has the following properties: (i) there exists a positive constant C such that whenever $\mathrm{Im}\, z < 0$,

$$|F_c(z)| < C,$$

(ii) at each point on the half plane $\mathrm{Im}\, z < 0$,

$$F_c(z) = \lim_{\lambda \to \infty} \int_{-\lambda}^{\lambda} \frac{F(u)}{2\pi \mathrm{i}(z - u)} \mathrm{d}u \tag{10.19}$$

where F is the ordinary transform of f, and where $\lim_{\lambda\to\infty}\int_{-\lambda}^{\lambda}$ may be replaced by $\int_{-\infty}^{\infty}$ whenever $1<p\leqslant 2$, (iii)

$$\lim_{\alpha\to 0^-} F_c(y+i\alpha)=F(y)$$

as a limit in the mean of index $q=p/(p-1)$ when $1<p\leqslant 2$, and uniformly to the continuous transform F when $p=1$, (iv) the real and imaginary parts of F are related as in theorem 8.35 equations (8.64)–(8.67). □

Theorem 10.13 Suppose that $\Phi(z)$ is analytic on the half plane $\operatorname{Im} z<0$, and suppose that for some $C>0$ and some p satisfying $1<p<\infty$,

$$\int_{-\infty}^{\infty}|\Phi(y+i\alpha)|^p\,dy<C$$

whenever $\alpha<0$: then (i) there exists a function $F(y)$ in $L^p(-\infty,\infty)$ such that

$$\underset{\alpha\to 0^-}{\text{l.i.m.}(p)}\Phi(y+i\alpha)=F(y),$$

and also pointwise at a.a.y, (ii) the real and imaginary parts X and Y of F are related by (8.64) and (8.65), (iii) whenever $1<p\leqslant 2$ then Φ is equal to the complex Fourier transform of a causal function $f\in L^q$, $q=p/(p-1)$, whose transform is F. □

10.4 Positive-definite and distribution functions

The *positive-definite* functions were so-called by Bochner because they arise as a natural generalization of the class of functions whose Fourier transforms are positive and integrable. The *distribution functions* are so-called because they arise in the theories of probability distributions and of spectral distributions. The distribution functions are the easier to define and we start with these.

Definition A real valued function $S(y)$ defined on $(-\infty,\infty)$ is a distribution function if $|S(y)|$ is bounded on $(-\infty,\infty)$, if $S(y)$ is monotonically increasing, and if for every y, $S(y)=\frac{1}{2}[S(y^+)+S(y^-)]$. □

For example, the functions $S(y)=\operatorname{sgn} y$ and $S(y)=(1-e^{-|y|})\operatorname{sgn} y$ are distribution functions when $S(0)$ is put equal to zero in each case (but not otherwise).

A distribution function will not necessarily be absolutely continuous, but when it is then the derivative S', defined a.e., will be nowhere negative and will be in class $L(-\infty,\infty)$. In this case we can form a continuous function $f(x)$ as the inverse Fourier transform of S' and it will follow that f and S will be related directly through the following equations, at all y and x, respectively, when the constant A is put equal to $S(0)$:

$$S(y)=A+\lim_{\lambda\to\infty}\int_{-\lambda}^{\lambda} f(x)\left[\frac{\exp(-2\pi ixy)-1}{-2\pi ix}\right]dx \tag{10.20}$$

$$f(x)=\lim_{\lambda\to\infty}\int_{-\lambda}^{\lambda}\exp(2\pi ixy)dS(y). \tag{10.21}$$

The integral in (10.21) is a Riemann–Stieltjes integral, as described for instance in Apostol (1974).

The crucial point now is that there may still exist a continuous function $f(x)$ such that (10.20) and (10.21) remain valid everywhere even when the distribution S is not absolutely continuous, and when f possesses no locally integrable transform. Bochner found that the class of functions so created is identical to the class of positive-definite functions defined as follows.

Definition We call a real or complex valued function $f(x)$ positive-definite if (i) it is defined, bounded, and continuous on $(-\infty,\infty)$, and (ii) at all x $f^*(-x)=f(x)$, and (iii) for any points $x_1, x_2, \ldots, x_N$, $(N=1,2,3,\ldots)$ and any (complex) numbers $a_1, a_2, \ldots, a_N$

$$\sum_{m=1}^{N}\sum_{n=1}^{N} f(x_m-x_n)a_m a_n^* \geqslant 0. \quad \square$$

This test is not easy to apply, and in practice such functions are often recognized either through the sufficient condition of being the transform of a positive function in L, or through the following Bochner theorem relating them to distribution functions.

Theorem 10.14 For each distribution function $S(y)$ there will exist a positive-definite function $f(x)$ defined everywhere by (10.21) and it will follow that (10.20) holds everywhere when A is put equal to $S(0)$: conversely for every positive-definite function $f(x)$ the function $S(y)$ defined, for arbitrary real A, by (10.20) at all y will be a distribution function and (10.21) will be true at all x irrespective of the value of A. $\square$

As an example the distribution functions S_1 and S_2 defined by $S_1(y)=\frac{1}{2}\operatorname{sgn} y$, $S_1(0)=0$, and

$$S_2(y)=\begin{cases}1 & y>0\\ \frac{1}{2} & y=0\\ 0 & y<0\end{cases}$$

are both linked via (10.20) and (10.21) with the positive-definite function $f(x)=1$. This arbitrariness in respect of an additive constant on the distribution is analogous to that arising with indefinite integrals. The functions $\cos x$ and $\exp(-|x|)$ are both positive-definite, whilst the function $\sin x$ is not positive-definite.

We conclude with two results which follow from the above. First, if S is any distribution, and f is defined through (10.21) then the limits $S(\infty)$ and $S(-\infty)$ are defined and

$$f(0)=S(\infty)-S(-\infty).$$

Second if f is a continuous function with a non-negative transform $F\in L$, then at each point of continuity of $S(y)$, and so at a.a.y,

$$S(y)=\int_0^y F(u)\mathrm{d}u$$

where $S(y)$ is defined by (10.20) with $A=0$.

Further properties of distribution functions arise from the fact that they are monotonic, see sections (11.3) and (11.4).

11

Power spectra and Wiener's theorems

11.1 Introduction

In this chapter we describe a set of theorems which are particularly associated with the work of N. Wiener (1894–1964) (Wiener, 1930, 1951, 1964). The theorems apply to any function f which is L^2_{LOC} and is also such that

$$\lim_{X\to\infty} \frac{1}{2X} \int_{-X}^{X} f^*(x')f(x'+x)\mathrm{d}x' \tag{11.1}$$

converges for all x. It is quite remarkable that so much follows from this single condition. Wiener used the letter S to represent the class of such functions, but we will use the letter W (since S is often used for the class of good functions). Any function in L^2_{LOC} that is periodic will be in class W, but importantly many aperiodic functions whose graphs appear devoid of order are also in class W and the theory is useful in describing physical systems involving random noise.

If f belongs to W then the function of x defined everywhere by (11.1) is written $R_f(x)$. R_f is called the *autocorrelation* of f, since R_f plays a role for functions in W analogous to that played by ρ_{ff}, (3.3), for a function $f \in L^2$. The theory introduces two other functions, a *spectral function* S_f (the spectrum of f) and a *spectral density function* P_f, and these come to be related in a formal way as follows:

$$R_f \leftrightarrow P_f \tag{11.2}$$

$$S_f(y) = \int_0^y P_f(u)\mathrm{d}u \tag{11.3}$$

$$P_f = S'_f. \tag{11.4}$$

In electrical theory the variables x and y often correspond to time and frequency, respectively, and the function $f(x)$ may represent a voltage or current signal: when this is the case $P_f(y)$ represents the average power transfer per unit frequency at frequency y, whilst $S_f(y)$ is related to the total average power transfer at frequencies between zero and y. For this reason P_f is sometimes called the power density spectrum of f, whilst S_f is called the power spectrum or integrated power spectrum of f.

The formal scheme in (11.2)–(11.4) implies that $S_f(y)$ is an absolutely continuous function of y with a derivative equal a.e. to $P_f(y)$, and this is indeed often the case when f is aperiodic. However, in many instances, including all cases in which $f \in W$ is periodic (and non-zero), it turns out that P_f cannot be defined as an ordinary locally integrable function, and the scheme in (11.2)–(11.4) is only maintained if P_f is represented by a generalized function. Nevertheless, despite these complications, it is *always* possible to define S_f as an ordinary locally integrable function whenever f belongs to W. For this reason Wiener's treatment concentrates more on the function S_f than on P_f, and in this chapter we will avoid the use of delta functions and other generalized functions.

A function $f \in W$ will not necessarily possess an ordinary locally integrable Fourier transform, but instead one can form the transform, F_X, of a truncated version of f as follows, for some positive X,

$$F_X(y) = \int_{-X}^{+X} f(x) \exp(-2\pi i x y) \mathrm{d}x. \tag{11.5}$$

An important part of Wiener's theory consists in showing that S_f and P_f may be related to F_X. When P_f is defined this relation may be expressed formally as

$$P_f(y) = \lim_{X \to \infty} \frac{1}{2X} |F_X(y)|^2: \tag{11.6}$$

it is important to realize, however, that the limit will not necessarily exist as a simple p.w.a.e. limit. Equation (11.6) can be made rigorous in various ways which we return to in theorem 11.13 and in (11.34).

Two different approaches to this subject are found in the literature. In one approach $P_f(y)$ is defined by a limit similar to that in (11.6), and then the fact that R_f and P_f form a Fourier pair is referred to as Wiener's theorem. In the other approach P_f is defined as the Fourier transform of R_f, and then a relation such as (11.6) is referred to as Wiener's theorem. Wiener himself adopted an approach closer to the latter, though his approach was to devise a way of defining S_f directly from R_f, and then to define P_f as the derivative of S_f whenever S_f is absolutely continuous. Relations between P_f and F_X then follow as theorems.

A further development of these ideas occurs in the work of both Wiener and of A. I. Khintchine (1894–1959) on stationary random processes (or stochastic processes) – that is for ensembles of functions governed by certain probability distributions. In Khintchine's approach an autocorrelation function is defined as $\langle f^*(x')f(x'+x)\rangle$, where the braces $\langle\ \rangle$ indicate an ensemble average of the enclosed quantity at fixed x' and x. In a stationary random process this quantity depends only on x, and not on x', and the resulting quantity plays a role in the theory of random processes very close to that of R_f defined by (11.1). Indeed relations such as (11.2)–(11.4) and (11.6) have their counterparts in stochastic theory, and in contexts in which it is not important to distinguish the two approaches the theory is often referred to as the Wiener–Khintchine theory. It is important to realize, however, that Wiener's basic theory of 'generalized harmonic analysis' is in no way probabilistic, and the theorems apply to single well defined functions rather than to ensembles of functions. The material summarized in this chapter makes no use of the theory of probability.

11.2 The autocorrelation function

We have already defined, in (11.1), the autocorrelation function R_f of a function $f\in W$: however, there are various other limits which also yield R_f. In the following theorem we show that the choice of origin is unimportant, and also that R_f can be defined by means of various truncated versions of f.

Theorem 11.1 Suppose $f\in W$ has an autocorrelation function R_f: then

(i) for any real a and all x,

$$R_f(x)=\lim_{X\to\infty}\frac{1}{2X}\int_{a-X}^{a+X} f^*(x')f(x'+x)\mathrm{d}x';$$

(ii) if we define f_X, for $X>0$, by

$$f_X(x)=\begin{cases} f(x), & \text{a.e. on } [-X,X] \\ 0, & |x|>X \end{cases} \tag{11.7}$$

then at all x,

$$R_f(x)=\lim_{X\to\infty}\frac{1}{2X}\int_{-\infty}^{+\infty} f_X^*(x')f_X(x'+x)\mathrm{d}x';$$

moreover this limit is bounded over $x\in(-\infty,\infty)$ and X greater than some X_0, and the limit converges as a l.i.m.(2) over any finite interval of x;

(iii) the product f sinc belongs to L^2, where the sinc function is defined by

$$\operatorname{sinc} x=\frac{\sin x}{x}\quad(=1 \text{ at } x=0); \tag{11.8}$$

(iv) at all x,

$$R_f(x) = \lim_{a \to \infty} \frac{2}{a} \int_{-\infty}^{\infty} f^*(x') \operatorname{sinc}(\pi x'/a) f(x'+x) \operatorname{sinc}[\pi(x'+x)/a] \mathrm{d}x'. \quad \square$$

The autocorrelation, R_f, has the following properties.

Theorem 11.2 Suppose $f \in W$ has autocorrelation function R_f: then

(i) $R_f(0)$ is real and equal to the mean square modulus of f, that is,

$$R_f(0) = \lim_{X \to \infty} \frac{1}{2X} \int_{-X}^{X} |f(x)|^2 \mathrm{d}x; \tag{11.9}$$

(ii) at all x,

$$R_f(-x) = [R_f(x)]^*; \tag{11.10}$$

(iii) at all x

$$|R_f(x)| \leqslant R_f(0); \tag{11.11}$$

(iv) R_f is not necessarily continuous, but it differs from a continuous function on at most a null set;

(v) whenever $R_f(x)$ is continuous at $x=0$ then it is continuous at all x;

(vi) R_f is positive-definite. □

The definition and some properties of a positive-definite function have been given in section 10.4.

We now give examples of functions in class W of various types. A function $f \in L^2_{\mathrm{LOC}}$ which is periodic will necessarily belong to W, and its autocorrelation function will also be periodic. Thus, trivially, the function $f(x)=1$ has $R_f(x)=1$ at all x. The functions $\cos 2\pi x$ and $\sin 2\pi x$ are each in class W, and each has the autocorrelation $R_f(x)=(\frac{1}{2})\cos 2\pi x$; this shows that different functions may share the same autocorrelation function. The function $\exp(2\pi \mathrm{i} x)$ has the autocorrelation function $R_f(x)=\exp(2\pi \mathrm{i} x)$, thus showing, when compared with the previous examples, that the autocorrelation of the sum of two functions is not necessarily equal to the sum of the autocorrelations.

In contrast to the periodic functions there is a class of functions in W whose autocorrelation, $R_f(x)$, tends to zero as $x \to \pm\infty$, R_f having a Fourier transform in L. These functions provide a model for the functions that arise in physical systems with noise; such a function is sometimes called a pseudo-random function. The following step function is of this type. We construct f so that $f(x)=f(-x)$ at all x, and so that on each interval $[n, n+1)$ of the x-axis, for $n=0, 1, 2, 3, \ldots$, $f(x)$ has either the value $+1$ or -1, the sign being as in the following sequence starting at $n=0$:

$$\left.\begin{array}{l} +,\ -; \\ +,\ +;\ +,\ -;\ -,\ +;\ -,\ -;\ \text{repeated twice} \\ +,\ +,\ +;\ +,\ +,\ -;\ +,\ -,\ +;\ +,\ -,\ -;\ -,\ +,\ +; \\ \quad -,\ +,\ -;\ -,\ -,\ +;\ -,\ -,\ -;\ \text{repeated four times} \\ +,\ +,\ +,\ +;\ +,\ +,\ +,\ -;\ +,\ +,\ -,\ +;\ \text{etc repeated} \\ \quad \text{eight times, and so on.} \end{array}\right\} \tag{11.12}$$

This function has the autocorrelation function

$$R_f(x)=\begin{cases}1-|x|, & -1<x<1\\ 0, & |x|\geqslant 1.\end{cases}$$

We pass now to examples of functions in W which are not commonly discussed in the context of physical applications. Consider, for instance, the following step function, which is defined just as above except that the sequence of signs in (11.12) is replaced by the following sequence:

$$\left.\begin{array}{l} +;\ -;\ -,\ +;\ -,\ +,\ +,\ -;\ -,\ +,\ +,\ -,\ +,\ -,\ -,\ +; \\ -,\ +,\ +,\ -,\ +,\ -,\ -,\ +,\ldots,\text{etc} \end{array}\right\}. \tag{11.13}$$

The reader will notice that the sequence of signs between successive semicolons is obtained from the complete preceding set of signs simply by replacing + by − and − by +. The resulting step function is aperiodic and has an autocorrelation function that is neither periodic nor tends to zero as $x\to\pm\infty$; $R_f(x)$ is a complicated but continuous function of x returning to the value 1/3 when $x=3,6,12,24,48,\ldots$ and to the value $-1/3$ when $x=1$, 2, 4, 8, 16, 32, 64, ...

We return to the classification of functions in W in section 11.3, but conclude with the following examples. Rather trivially any function $f\in L^2(-\infty,\infty)$ will be in W, but it yields $R_f(x)=0$ at all x. The function $\exp(\mathrm{i}|x|^{1/2})$ is in W and leads to the autocorrelation $R_f(x)=1$ at all x, thus having the same autocorrelation as the function $f(x)=1$. The functions $\sin(x^2)$, $\cos(x^2)$, are both in W and share the autocorrelation

$$R_f(x)=\begin{cases}\tfrac{1}{2}, & x=0\\ 0, & x\neq 0.\end{cases} \tag{11.14}$$

Although some functions in W, such as $\sin x^2$, see (9.5), do possess ordinary locally integrable Fourier transforms, such functions always have $R_f(x)=0$ for a.a.x, and are of restricted interest.

11.3 The spectrum and spectral density

We have seen that a function in W will not necessarily possess an ordinary locally integrable Fourier transform, nor will its autocorrelation function. However, one can always define a function S_f called the spectrum of f, on account of the following theorem.

Theorem 11.3 Suppose $f \in W$ has autocorrelation R_f: then

$$\lim_{\lambda \to \infty} \int_{-\lambda}^{\lambda} R_f(x) \left\{ \frac{\exp(-2\pi i x y) - 1}{-2\pi i x} \right\} dx \tag{11.15}$$

converges at all y to define a function $S_f(y)$ which we call the spectrum of f. □

Alternative names for S_f are 'power spectrum' or 'integrated spectrum' or 'integrated power spectrum'.

The spectrum S_f is not merely a locally integrable function but has the following properties which make it a distribution function as defined in section 10.4.

Theorem 11.4 Suppose $f \in W$ has autocorrelation R_f and spectrum S_f: then

(i) $S_f(y)$ is real valued for all y and increases monotonically with y, i.e. $S_f(b) \geqslant S_f(a)$ whenever $b > a$;

(ii) at all y, $S_f(y) = \frac{1}{2}[S_f(y^+) + S_f(y^-)]$;

(iii) at each $\varepsilon > 0$ the function $[S_f(y+\varepsilon) - S_f(y-\varepsilon)]$ is L^2 on $-\infty < y < \infty$;

(iv) $S_f(0) = 0$;

(v) the limits $S_f(\infty)$ and $S_f(-\infty)$ exist, and

$$S_f(\infty) - S_f(-\infty) \leqslant R_f(0),$$

and whenever $R_f(x)$ is continuous at $x = 0$ then

$$S_f(\infty) - S_f(-\infty) = R_f(0). \quad \square$$

These properties appear in the following examples. The function $\sin 2\pi x$ has autocorrelation function $\frac{1}{2}\cos 2\pi x$ and its spectrum S_f is the following step function:

$$S_f(y) = \begin{cases} -1/4 & y < -1 \\ -1/8 & y = -1 \\ 0 & -1 < y < 1 \\ +1/8 & y = 1 \\ +1/4 & y > 1. \end{cases}$$

In contrast to this discontinuous spectrum, the step function described through (11.12) has the following continuous spectrum:

$$S_f(y) = \int_0^y \left[\frac{\sin(\pi u)}{(\pi u)} \right]^2 du. \tag{11.16}$$

The formula in (11.15) is reminiscent of that giving the integral of the transform of a function, see (7.7) and theorem 10.3, and in those cases when R_f possess an ordinary transform we may use the following.

Theorem 11.5 Suppose $f \in W$ has autocorrelation R_f and spectrum S_f: then R_f will possess a Fourier transform in L_{LOC} if, and only if, S_f is an absolutely continuous function, and when this is so we define the spectral density P_f of f as equal to the derivative $\mathrm{d}S_f/\mathrm{d}y$ at all points at which the derivative exists; moreover the function P_f so defined has the following properties:

(i) $P_f(y) \geqslant 0$ a.e.,
(ii) $P_f \in L$,
(iii) at all y, $S_f(y) = \int_0^y P_f(u)\mathrm{d}u$,
(iv) $R_f \leftrightarrow P_f$, and

$$\int_{-\infty}^{+\infty} P_f(y) \exp(2\pi \mathrm{i} xy)\mathrm{d}y$$

defines a continuous function equal a.e. to $R_f(x)$. □

Part (iv) of the above theorem gives an inverse to (11.15), showing how to regain R_f from S_f when S_f is absolutely continuous. For completeness we give a general formula applicable for all functions $f \in W$: the formula is based on the so-called Riemann–Stieltjes integral, and for the interpretation of the symbol $\mathrm{d}S_f(y)$ we refer the reader, for instance, to Apostol (1974).

Theorem 11.6 Suppose $f \in W$ has autocorrelation R_f and spectrum S_f: then

$$\lim_{Y \to \infty} \int_{-Y}^{Y} \exp(2\pi \mathrm{i} xy)\mathrm{d}S_f(y) \tag{11.17}$$

converges at all x to a continuous function equal a.e. to $R_f(x)$. □

We have given examples of functions in which the spectrum was, respectively, a step function and an absolutely continuous function. This suggests that functions in W may conveniently be classified according to the properties of the spectrum. The fact that S_f is a monotonically increasing function allows us to use the following.

Theorem 11.7 Any monotonically increasing function defined everywhere on the real line can be expressed as the sum of three monotonically increasing functions, (*a*) a step function having a finite or denumerably infinite set of step discontinuities, and (*b*) an absolutely continuous function, and (*c*) a continuous function with a derivative equal a.e. to zero; moreover the functions in (*a*), (*b*) and (*c*) are unique apart from additive constants. □

Thus a spectrum $S_f(y)$ may be decomposed uniquely into three component spectra of types (*a*), (*b*) and (*c*), respectively, say $S_a(y)$, $S_b(y)$ and $S_c(y)$, each being chosen to have the value zero at $y=0$.

The stepped portion, S_a, is determined entirely by the discontinuities in S_f and these may in turn be determined from R_f using the following relation.

Theorem 11.8 Let $f \in W$ have autocorrelation R_f and spectrum S_f: then at all y

$$\lim_{X\to\infty} \frac{1}{2X}\int_{-X}^{X} R_f(x)\exp(-2\pi i x y)\mathrm{d}x = S_f(y^+) - S_f(y^-). \quad \square \quad (11.18)$$

The two sides of (11.18) will equal zero at a.a.y. On subtracting S_a from S_f the remainder is differentiable a.e. to give the derivative $S_b' + S_c'$. The absolutely continuous portion is then given at all y by

$$S_b(y) = \int_0^y [S_b'(y) + S_c'(y)]\mathrm{d}y.$$

S_c is then obtained by subtraction, completing the decomposition.

In the next sections we discuss the three types of spectrum in turn.

11.4 Discrete spectra

A function $f \in W$ whose spectrum $S_f(y)$ is a step function with a finite or a denumerably infinite set of steps is said to possess a *discrete spectrum* (or a *pure line spectrum*). The spectrum is in this case characterized entirely by the coordinates $y_1, y_2, y_3, \ldots$ of the steps and by the positive numbers $k_1, k_2, k_3, \ldots$ representing the step heights

$$k_n = S_f(y_n^+) - S_f(y_n^-).$$

Indeed for any real numbers a and b, $b > a$, we have

$$S_f(b) - S_f(a) = \sum_{n=1}^{\infty} k_n g_n, \qquad S_f(0) = 0, \qquad (11.19)$$

where

$$g_n = \begin{cases} 1, & y_n \in (a,b) \\ \tfrac{1}{2}, & y_n = a \text{ or } y_n = b \\ 0, & y_n \notin [a,b]. \end{cases}$$

In (11.19) we have used the infinite sum to cover the case of a denumerably infinite set of steps, but in this and other cases in this section it is understood that this is to be replaced by a finite sum when the number of steps is finite. It is convenient to borrow terminology from the language of optics and refer to the k_n as line intensities, and to the y_n as line frequencies or spectral frequencies.

A well known special case occurs when each of the spectral frequencies y_n can be expressed as an integer multiple (positive, negative or zero) of some number $1/\tau$, $\tau>0$. In this case at a.a.x:

$$R_f(x)=\sum_{n=1}^{\infty} k_n \exp(2\pi i x y_n) \tag{11.20}$$

where the summation converges uniformly over $-\infty<x<\infty$ to a continuous function of period τ. The k^n are in this case the Fourier series coefficients of R_f:

$$k_n=\frac{1}{\tau}\int_0^{\tau} R_f(x)\exp(-2\pi i x y_n)\mathrm{d}x.$$

Note that in (11.20) the index n merely labels the y_n, not necessarily in any special order, so that the sum runs from $n=1$ to $n=\infty$ rather than from $-\infty$ to $+\infty$.

Whenever f is L^2_{LOC} and of period τ the above conditions are met, and $R_f(x)$ will then be a continuous function of period τ; we can in this case invoke the Fourier series coefficients F_n of f and show that for $n=1,2,3,\ldots$

$$k_n=|F_n|^2$$

and

$$\tau^{-1}\int_0^{\tau}|f(x)|^2\mathrm{d}x=R_f(0)=\sum_{n=1}^{\infty}|F_n|^2 \tag{11.21}$$

where

$$F_n=\tau^{-1}\int_0^{\tau} f(x)\exp(-2\pi i x y_n)\mathrm{d}x.$$

Equation (11.21) is a form of *Parseval formula* for periodic functions in L^2_{LOC}; see also section 15.7.

The above results can be generalized to cover any type of discrete spectrum, and the class of *almost periodic functions* arises as a generalization of the class of continuous periodic functions. The definition of an almost periodic function relies on defining first a quantity τ_ε which is a generalization of the concept of period.

Definition Suppose $f(x)$ is a continuous, real or complex valued function defined on $-\infty<x<\infty$, and consider some $\varepsilon>0$; any number τ_ε such that at all x

$$|f(x+\tau_\varepsilon)-f(x)|\leqslant\varepsilon$$

is called a *translation number of* $f(x)$ belonging to ε: if in addition $f(x)$ is such that for every $\varepsilon>0$ there exists a real number L_ε such that no interval

$(a, a+L_\varepsilon)$ is free of translation numbers belonging to ε then $f(x)$ is said to be almost periodic. □

A simple example of an almost periodic function is provided by $(\cos 2x + \cos 2^{1/2}x)$. This function displays beats between the two component cosine functions, but, because the numbers 2 and $2^{1/2}$ are not commensurate the function never exactly repeats itself. Nevertheless values of τ_ε and L_ε exist for every $\varepsilon>0$, though as ε is reduced so the minimum value of L_ε becomes larger. A continuous periodic function of period τ is a special case of an almost periodic function, the numbers $\tau, 2\tau, 3\tau, \ldots$ being translation numbers for every $\varepsilon>0$.

The relevance of almost periodic functions to discrete spectra is as follows.

Theorem 11.9 Suppose $f \in W$ has autocorrelation function R_f and a discrete spectrum $S_f(y)$ with steps k_n at positions y_n, $n=1,2,3,\ldots$: then

(i) at a.a.x

$$R_f(x) = \sum_{n=1}^{\infty} k_n \exp(2\pi i x y_n) \tag{11.22}$$

where the sum converges uniformly on $-\infty < x < \infty$ to an almost periodic function (differing from R_f on a null set),

(ii) at each n,

$$k_n = \langle R_f(x) \exp(-2\pi i x y_n) \rangle_x,$$

(iii)

$$\langle |f(x)|^2 \rangle_x \geqslant \sum_{n=1}^{\infty} k_n,$$

(iv) f is not necessarily almost periodic. □

In the above and subsequently we use the symbol $\langle \ldots \rangle_x$ to represent an average value, over all values of x, of the enclosed function of x; thus, by definition,

$$\langle g(x) \rangle_x = \lim_{X \to \infty} \int_{-X}^{X} \frac{g(x)}{2X} \, dx, \tag{11.23}$$

whenever this limit exists.

If in theorem 11.9 the function f is itself almost periodic then we can say more.

Theorem 11.10 Let f be an almost periodic function: then

(i) $f \in W$ and the autocorrelation function R_f is also almost periodic and the spectrum S_f will be discrete, so that the conditions of theorem 11.9 are satisfied and we may define the denumerable sets k_n and y_n;

(ii) (11.22) is valid at *all* x and R_f is continuous;

(iii) at each y_n we may define a complex number F_n by

$$F_n = \langle f(x)\exp(-2\pi i x y_n)\rangle_x \tag{11.24}$$

and it follows that

$$k_n = |F_n|^2;$$

(iv) if in (11.24) y_n is replaced by a real number other than one of the spectral frequencies then the average $\langle\ \rangle_x$ exists and equals zero;

(v) $$\langle |f(x)|^2\rangle_x = \sum_{n=1}^{\infty} k_n; \tag{11.25}$$

(vi) $$\lim_{N\to\infty} \langle |f(x) - \sum_{n=1}^{N} F_n \exp(2\pi i x y_n)|^2\rangle_x = 0. \quad \square \tag{11.26}$$

Notice that (11.25) is a version of Parseval's formula for almost periodic functions, and that the complex numbers F_n reduce to the familiar Fourier series coefficients of f when f is periodic. Equation (11.26) represents a form of mean square convergence.

It is important to note that although

$$\sum_{n=1}^{\infty} F_n \exp(2\pi i x y_n) \tag{11.27}$$

converges to $f(x)$ in the sense indicated in (11.26), it will not necessarily converge uniformly over $(-\infty, \infty)$ and indeed may fail to converge to $f(x)$ at certain points. The Fejér function (5.9), for example, is almost periodic yet (11.27) will fail to converge to $f(x)$ at $x=0$. Despite this fact, it is possible to approximate arbitrarily closely to an almost periodic function by means of a trigonometric polynomial; indeed this is the basis of the following fundamental characterization of almost periodic functions established by H. Bohr.

Theorem 11.11 A real or complex valued function $f(x)$, defined on $-\infty < x < \infty$, will be an almost periodic function if, and only if, for every $\varepsilon > 0$ there exists a finite set of complex numbers $A_1, A_2, A_3, \ldots, A_N$ and a set of real numbers $Y_1, Y_2, Y_3, \ldots, Y_N$, such that for all x

$$|f(x) - \sum_{n=1}^{N} A_n \exp(2\pi i x Y_N)| \leqslant \varepsilon. \quad \square \tag{11.28}$$

The Fejér function warns us that the values of the A_n and Y_n may have to be reassigned as ε is made smaller, and that they cannot necessarily be chosen simply as subsets of the F_n and y_n which belong to f.

11.5 Continuous spectra

The continuous part of any spectrum is itself the sum of an absolutely continuous component and a component whose differential coefficient is a.e. zero, and we consider first those functions for which the second component is zero. The function defined by (11.12) is of this type, and for such functions we have already seen how a spectral density can be defined and used in theorem 11.5.

Functions in W which possess a spectral density and have an absolutely continuous spectrum are important because in a certain sense they are the most common type of function in W, and in applications involving random noise such functions are the most likely to arise. In the following theorem this statement is made precise for a particular type of step function; the idea is that each different step function is associated with a different number in the range [0, 1), and it is shown that for almost all such numbers the corresponding step function has an integrable spectral density.

Theorem 11.12 Consider any step function of the form

$$f(x)=\begin{cases}2a_n-1, & n-1\leqslant x<n\\ 2b_n-1, & -n\leqslant x<-n+1\end{cases}$$

where, for $n=1,2,3,\ldots$, each a_n and each b_n has a value of 0 or $+1$, and exclude those functions for which both a_n and b_n become unity for n greater than some N; let p be the number whose representation in binary notation is

$$p=0.a_1b_1a_2b_2a_3b_3\ldots:$$

then it follows that (i) to each such step function there corresponds a unique number $p\in[0,1)$ and vice versa, (ii) for almost all $p\in[0,1)$ the corresponding step function has autocorrelation and spectral density

$$R_f(x)=\begin{cases}1-|x|, & -1<x<1\\ 0, & |x|\geqslant 1\end{cases}$$

$$P_f(y)=\left(\frac{\sin \pi y}{\pi y}\right)^2. \quad \square$$

We now consider the cases in which the spectrum is continuous and has a differential coefficient a.e. of zero. The example in (11.13) is of this type. It might be thought that a continuous function with a zero differential a.e. must be equal simply to a constant, but the following counterexample (based on Cantor's ternary set) shows that this is not so. Construct a function $g(x)$ so that $g(0)=0$ $(x\leqslant 0)$ and $g(1)=1$ $(x\geqslant 1)$; and on each closed interval whose end points, in ternary notation, are of the form

$$x=0.a_1a_2a_3\ldots a_m\,0\,2\,2\,2\,2\ldots$$

$$x=0.a_1a_2a_3\ldots a_m\,2\,0\,0\,0\,0\ldots,$$

where each a_n is either zero or 2, assign to $g(x)$ the value, in binary notation,

$$g(x)=0.b_1b_2b_3\ldots b_m\ 1\,0\,0\,0\,0\ldots$$

where $b_n=a_n/2$. In this way $g(x)$ is defined a.e. on $[0, 1]$ and it is possible to complete the definition in such a way that for all $x\in[0, 1]$

$$g(x)=\lim_{\varepsilon\to 0}\frac{1}{2\varepsilon}\int_{x-\varepsilon}^{x+\varepsilon} g(u)\mathrm{d}u.$$

The resulting function g is continuous on $[0, 1]$, has a differential coefficient of zero a.e. on $[0, 1]$ and yet $g(x)$ increases monotonically from $g(0)=0$ to $g(1)=1$. For example, $g(x)$ equals 1/2 on $[1/3, 2/3]$ and 1/4 on $[1/9, 2/9]$.

11.6 Miscellaneous theorems

We describe first some theorems relating to the truncation of a function in class W. These theorems lead to well known expressions for the spectrum and spectral density.

Given a function $f\in W$ we consider a portion of the x-axis of length $2X$ $(X>0)$ centred on the point $x=\alpha$, for any real α, and define functions f_X, F_X and ρ_X as follows:

$$f_X(\alpha, x)=\begin{cases} f(x), & \text{a.a.}x \text{ on } [\alpha-X, \alpha+X] \\ 0, & |x-\alpha|>X, \end{cases} \tag{11.29}$$

$$F_X(\alpha, y)=\int_{\alpha-X}^{\alpha+X} f(x)\exp(-2\pi\mathrm{i}xy)\mathrm{d}x, \tag{11.30}$$

$$\rho_X(\alpha, X)=\int_{-\infty}^{+\infty} f_X^*(\alpha, x')f_X(\alpha, x'+x)\mathrm{d}x'. \tag{11.31}$$

F_X and ρ_X are thus the Fourier transform and autocorrelation function of the truncated function f_X. There are senses in which $|F_X(\alpha, y)|^2/2X$ tends to the spectral density $P_f(y)$ as $X\to\infty$, but it is important to realize that this will *not* necessarily occur as a simple pointwise limit even when P_f is everywhere continuous and infinitely differentiable. For instance if we take f to be as defined by (11.12), then we find that $P_f(0)=1$ but that as $X\to\infty$ so $|F_X(0, 0)|^2/2X$ oscillates finitely above and below the value of unity without converging. One can achieve convergence by performing a local average over a range of values of y *or* by taking a running average over all values of α, as follows.

Theorem 11.13 Suppose $f\in W$ has a spectral density $P_f\in L$, and let F_X be as in (11.30): then

(i) for any fixed value of α, at a.a.y, and at each point of continuity of

$P_f(y)$

$$\lim_{\varepsilon\to 0}\lim_{X\to\infty}\frac{1}{2\varepsilon}\int_{y-\varepsilon}^{y+\varepsilon}\frac{|F_X(\alpha,u)|^2}{2X}\,du=P_f(y),$$

(ii) at a.a.y and at each point of continuity of $P_f(y)$

$$\lim_{X\to\infty}\lim_{A\to\infty}\frac{1}{2A}\int_{-A}^{A}\frac{|F_X(\alpha,y)|^2}{2X}\,d\alpha=P_f(y)$$

(iii) at any fixed value of X, and uniformly over $-\infty<y<\infty$,

$$\lim_{A\to\infty}\frac{1}{2A}\int_{-A}^{A}\frac{|F_X(\alpha,y)|^2}{2X}\,d\alpha=P_f*K_X(y)$$

where

$$K_X(y)=2X\left(\frac{\sin 2\pi Xy}{2\pi Xy}\right)^2,$$

(iv) if a function $g\in L$ has transform G, then at each α

$$\lim_{X\to\infty}\int_{-\infty}^{+\infty}\frac{|F_X(\alpha,y)|^2}{2X}G(y)dy=\int_{-\infty}^{\infty}P_f(y)G(y)dy.\quad\square$$

If we drop the condition that f has a spectral density then the following results remain.

Theorem 11.14 Suppose $f\in W$ has spectrum S_f and autocorrelation R_f and let F_X and ρ_X be as in (11.30) and (11.31): then

(i) for each α, at a.a.y and at every point of continuity of $S_f(y)$,

$$\lim_{X\to\infty}\int_0^y\frac{|F_X(\alpha,u)|^2}{2X}\,du=S_f(y);$$

(ii) at each α, and at each x, including points of discontinuity of $R_f(x)$,

$$\lim_{X\to\infty}\frac{\rho_X(\alpha,x)}{2X}=R_f(x),$$

where at fixed α the convergence is uniformly bounded over $x\in(-\infty,\infty)$ for $X>1$;

(iii) at each X and at each x,

$$\lim_{A\to\infty}\frac{1}{2A}\int_{-A}^{A}\frac{\rho_X(\alpha,x)}{2X}\,d\alpha=\begin{cases}R_f(x)\left(1-\dfrac{|x|}{2X}\right), & |x|<2X\\ 0, & |x|\geqslant 2X\end{cases}$$

where at fixed X the convergence is uniformly bounded on $-\infty<x<\infty$ and $A>1$;

(iv) at each value of y

$$\lim_{A\to\infty}\frac{1}{2A}\int_{-A}^{A}\frac{|F_X(\alpha, y)|^2}{2X}\,d\alpha = \lim_{Y\to\infty}\int_{-Y}^{Y}\frac{\sin^2[2\pi X(y'-y)]}{2X\pi^2(y'-y)^2}\,dS_f(y') \tag{11.32}$$

and the limit function is a continuous, bounded, non-negative, integrable function of y. □

The above theorems were concerned with the product of the function f with a rectangle function, and in Wiener (1930, 1951, 1964) various analogous results are given for functions other than the rectangle function. We now turn to some theorems dealing with the convolution of f with some function. These results are relevant to the process of smoothing and filtering of signals.

Theorem 11.15 Consider a function $f \in W$ having a spectrum S_f, and let h be a function such that $(1+x)h(x)$ is both in L and L^2 on $-\infty < x < \infty$: then

(i) $fh \in L$ and the convolution $g = f * h$ is in class W with an autocorrelation R_g which is everywhere continuous;

(ii) whenever f possesses a spectral density in L so also will g, and if these densities are P_f and P_g then, at each y at which $P_f(y)$ is defined,

$$P_g(y) = P_f(y)|H(y)|^2, \tag{11.33}$$

where H is the continuous Fourier transform of h;

(iii) at all x

$$R_g(x) = R_f * \rho_{hh}(x)$$

where ρ_{hh} is the autocorrelation function of h, (3.3);

(iv) at each point of continuity of $S_g(y)$

$$S_g(y) = \int_0^y |H(u)|^2 dS_f(u). \quad \square$$

In concluding this chapter we point out that Wiener's results can be expressed very neatly if the formalism of generalized functions is used, chapters 12–13. We can then always associate a spectral density $\tilde{P}_f$ with a function $f \in W$, where $\tilde{P}_f$ is a generalized function in class S', and $\tilde{P}_f$ is equal to the generalized derivative of the ordinary function S_f. Functions with a discrete spectrum have a $\tilde{P}_f$ consisting of delta functions. Always we find $R_f \leftrightarrow \tilde{P}_f$ when $f \in W$.

Using the generalized spectral density $\tilde{P}_f$ it *is* valid to write

$$\tilde{P}_f = \lim_{X\to\infty}\frac{|F_X(\alpha, y)|^2}{2X}, \tag{11.34}$$

for each real α provided the limit is treated as a generalized limit in S'. Also, for example, (11.32) can be replaced, at each y, by

$$\left\langle \frac{|F_X(\alpha, y)|^2}{2X} \right\rangle_\alpha = \tilde{P}_f * K_X(y)$$

where the left hand limit is an ordinary one, and the convolution is a generalized convolution. By using $\tilde{P}_f$ instead of S_f the use of Stieltjes integrals may be avoided.

12

Generalized functions

12.1 Introduction

The Dirac delta function $\delta(x)$ is the best known of a class of entities called *generalized functions*. The generalized functions are important in Fourier theory because they allow any function in L_{LOC} (and indeed any generalized function also) to be Fourier transformed. Thus the function $f(x)=1$ has no Fourier transform within the realm of functions in L_{LOC}, but it acquires the transform $\delta(y)$ in the generalized theory. The generalized functions thus remove a blockage which existed in the previous theory. There is an analogy with the way in which the use of complex numbers allows any quadratic equation to be solved, whilst within the realm only of real numbers not all quadratic equations have solutions.

Generalized functions remove many other blockings which occur in the analysis of functions in L_{LOC}. For instance, *every* locally integrable function (and indeed every generalized function) can be regarded as the integral of some generalized function and thus becomes infinitely differentiable in this new sense. Many sequences of functions which do not converge in any accepted sense to a limit function in L_{LOC} can be regarded as converging to a generalized function, and moreover in this case the sequence of Fourier transforms will necessarily converge to the Fourier transform of the limit. Thus, in many ways the use of generalized functions simplifies the rules of analysis.

The Dirac delta function $\delta(x)$ is sometimes described as having the value zero for $x\neq 0$ and the value of infinity for $x=0$. This is a dangerous statement because it implies that a generalized function is specified by giving its value for all, or almost all, values of x. In fact they are specified in quite a different manner, and this emphasizes that they are in reality very different entities from functions in L_{LOC}.

A safer definition of the delta function considers it as the limit of a sequence of functions, for instance

$$\delta(x)=\lim_{n\to\infty} n\exp(-\pi n^2x^2).$$

This approach typifies one strand in the history of generalized functions, and Jones (1966) and Lighthill (1959) are based on this approach. Another definition of the delta function is based on the idea that if a function $f\in L_{\mathrm{LOC}}$ is continuous at $x=0$, then $\delta(x)$ is defined so that

$$\int_{-\infty}^{+\infty}\delta(x)f(x)\mathrm{d}x=f(0).$$

This approach typifies an alternative, and rather more formal, approach pioneered by Schwartz (1966), and employed in Gel'fand *et al.* (1964–9), Shilov (1968) and Zemanian (1965). In our description we will draw on both approaches.

In addition to these purely mathematical matters it is worth noting that the generalized functions are the natural mathematical entities with which to describe many of the abstractions which occur in physical theory. The impulsive force, the point charge, the point dipole, and the frequency response of an undamped harmonic oscillator are all aptly represented by generalized functions.

Generalized functions of several types have been established, and in the present chapter we describe those called functionals in S' or tempered distributions. In chapter 14 we extend the treatment to other types called functionals in D' (distributions) and functionals in Z' (ultradistributions). Several other classes of generalized function are described in Gel'fand *et al.* (1964–9).

12.2 The definition of functionals in S'

We start using an approach based on sequences of functions, and relying on the class S of good functions (chapter 7). Consider a sequence $\{f_n\}_{n=1}^{\infty}$ of ordinary functions such that $f_n\varphi\in L$ for each $\varphi\in S$ and each n, and suppose that

$$\lim_{n\to\infty}\int f_n(x)\varphi(x)\mathrm{d}x \tag{12.1}$$

converges for each $\varphi\in S$: the sequence $\{f_n\}$ is then said to *converge in S′*. It may be that there exists a function $f\in L_{\mathrm{LOC}}$ such that the value of the limit in (12.1) is, for each φ, equal to $\int_{-\infty}^{\infty} f\varphi$, and when this is the case the sequence is said to converge to f in S'. However, it may be that no such limit function exists even though the sequence converges in S'. This is the case for instance with the functions

$$f_n(x)=(n/2)\exp(-n|x|): \tag{12.2}$$

in this case the limit in (12.1) converges to the value $\varphi(0)$ but there is no ordinary limit function f such that $\varphi(0)=\int f\varphi$ for all $\varphi\in S$.

The generalized functions are now invented to play the role of a limit function irrespective of whether an ordinary limit function exists or not. If we note that the limiting process in (12.1) is in effect merely a way of associating a number with each φ in class S we arrive at the following definition of one type of generalized function.

Definition An association of exactly one real or complex number with each $\varphi\in S$ is said to be a *functional in class* S' if there exists at least one sequence of ordinary functions $\{f_n\}_{n=1}^{\infty}$ such that $f_n\varphi\in L$ for each $\varphi\in S$ and each n, and such that for each $\varphi\in S$ the number associated with φ is equal to $\lim_{n\to\infty}\int f_n\varphi$: any such sequence is said to converge in S' to the functional or to represent it. □

For example, the sequence of functions defined by (12.2) converges in S' and defines the functional in S' called the delta function (or, more properly, the delta functional).

It can be shown that each functional in S' can in fact be represented by a sequence $\{f_n\}$ of good functions, and the definition is sometimes based on the existence of at least one such sequence.

If the letter $\bar{f}$ or $\bar{f}(x)$ is used to represent a functional in S' then the symbol $\int \bar{f}\varphi$ or $\int_{-\infty}^{\infty}\bar{f}(x)\varphi(x)\mathrm{d}x$ is often used to represent the number that $\bar{f}$ associates with φ: the symbol is merely an elaborate composite symbol and does not represent an integral constructed in the Riemann or Lebesgue sense. The symbolic statement

$$\lim_{n\to\infty} f_n=\bar{f} \quad (\text{in } S') \tag{12.3}$$

means that the sequence $\{f_n\}$ of ordinary functions converges in S' to $\bar{f}$ as $n\to\infty$, which in turn means that for each $\varphi\in S$,

$$\lim_{n\to\infty}\int f_n(x)\varphi(x)\mathrm{d}x=\int \bar{f}\varphi. \tag{12.4}$$

The symbol $\bar{f}(x)$ is not to be taken as the value that $\bar{f}$ associates with a particular value of x, since this has not been given meaning: the symbol is a composite symbol standing for the functional $\bar{f}$ in a context in which a variable x is being used.

An alternative approach is based on another type of convergence applicable to sequences of good functions, and on the concept of a linear functional. We define convergence in S as follows.

Definition A sequence $\{\varphi_n\}_{n=1}^{\infty}$ of functions $\varphi_n\in S$ is said to *converge in* S if for each pair of non-negative integers k and l the following limit converges uniformly on $-\infty<x<\infty$:

$$\lim_{n\to\infty} x^k \frac{d^l\varphi_n(x)}{dx^l}, \tag{12.5}$$

it being understood that the zeroth derivative of φ_n is equal to φ_n. □

It can be shown that whenever such a sequence converges in S then there will necessarily exist a unique limit function $\varphi \in S$ to which φ_n converges uniformly, and the sequence is said to *converge to* φ *in* S. For example, the sequence

$$\varphi_n(x)=\exp[-(x-1/n)^2]$$

converges in S to $\exp(-x^2)$. However, the sequence

$$\varphi_n(x)=\cos nx \exp(-x^2)$$

does not converge in S, though it does converge in S' to the limit function $\varphi(x)=0$ (see theorem 3.14).

An association of exactly one real or complex number with each of a specified class of functions is often called a *functional*. The functionals in S', as defined earlier, are thus functionals, though not every functional is a functional in S'. If F represents a functional, then the symbol $F[\varphi]$ is commonly used to represent the number associated with the function φ. A functional F defined on the good functions is said to be *linear* if for any real or complex numbers α and β, and any two good functions φ and ψ,

$$F[\alpha\varphi+\beta\psi]=\alpha F[\varphi]+\beta F[\psi]. \tag{12.6}$$

A functional F defined on functions in S is said to be *continuous on* S if for every sequence $\{\varphi_n\}_{n=1}^{\infty}$ of functions that is convergent in S,

$$\lim_{n\to\infty} F[\varphi_n]=F[\varphi] \tag{12.7}$$

where φ is the limit in S of the sequence $\{\varphi_n\}$. For example, a functional F may be defined on S by

$$F[\varphi]=\int_{-\infty}^{\infty} |\varphi(x)|^2 dx.$$

This functional is continuous on S, but it is not linear. The following functional F on S is neither linear nor continuous on S:

$$F[\varphi]=\begin{cases} 0, & \text{when } \varphi(0)=0 \\ 1, & \text{when } \varphi(0)\neq 0. \end{cases}$$

However, the functional defined on S by

$$F[\varphi]=\varphi(0)$$

can be shown to be both linear and continuous on S, and it is in fact the delta functional already described.

We now have the following fundamental theorem.

Theorem 12.1 A functional F defined on S will be linear and continuous on S if, and only if, there exists at least one sequence $\{f_n\}_{n=1}^{\infty}$ of ordinary functions such that for each good function φ

$$F[\varphi]=\lim_{n\to\infty}\int f_n\varphi. \quad \square$$

Thus the class of functionals in S' as defined earlier is identical to the class of linear continuous functionals on S, and the names are interchangeable. The functionals in S' are also sometimes called *tempered distributions*.

A generalized function is often represented by a similar symbol to that for functions in L_{LOC}, such as f, g, $f(x)$ or $g(x)$. The corresponding number associated with the good function φ is commonly written in one of many ways such as $\int f\varphi$, $\int_{-\infty}^{\infty} f(x)\varphi(x)\mathrm{d}x$, $\langle f, \varphi\rangle$, (f, φ), $f[\varphi]$ or $f(\varphi)$. In this book we will use symbols such as $\tilde{f}$, $\tilde{g}$, $\tilde{F}$ or $\tilde{G}$ when we wish to emphasize that the symbol represents a generalized function rather than an ordinary function and will use $\langle\tilde{f}, \varphi\rangle$ to represent the number that $\tilde{f}$ associates with φ: we use the tilde primarily for pedagogical clarity, and it is not an accepted convention.

A functional $\tilde{f}\in S'$ is said to be *regular in* S' if there exists an ordinary function f such that for each $\varphi\in S$

$$\langle\tilde{f},\varphi\rangle=\int_{-\infty}^{\infty} f(x)\varphi(x)\mathrm{d}x, \tag{12.8}$$

and when this is the case we say that $\tilde{f}$ and f are equal or equivalent and write

$$\tilde{f}=f.$$

In this way each ordinary function in K, defined in section 9.2, defines a regular functional in S', so that functions in K can be regarded as special cases of functionals in S'. This is analogous to the way in which the real numbers are often regarded as a subset of the complex numbers. A generalized function that is not regular is defined as *singular*, so that, for example, the delta functional, which we now write as $\tilde{\delta}$, is a singular functional in S'.

The class S clearly plays a crucial role in setting up the functionals in S', and the functions in S are often referred to as *test functions* when used in this context. The class of functionals in S' and the class of functions in S are also paired together as *dual spaces* of each other. Other classes of test function lead to other classes of generalized function, and we return to two such in chapter 14.

12.3 Basic theorems

A host of definitions and theorems go towards setting up the rules of manipulation of generalized functions, and we now run through some of these. We go rather quickly because the results are exactly as one might hope, and are easy to remember. The following results are described in terms of generalized functions in class S', but the modifications for classes D' an Z' will be easily made in due course.

Two functionals $\tilde{f}$ and $\tilde{g}$ in S' are said to be equal if $\langle\tilde{f},\varphi\rangle=\langle\tilde{g},\varphi\rangle$ for each $\varphi\in S$, and $\tilde{f}$ is said to be equal zero if $\langle\tilde{f},\varphi\rangle=0$ for all $\varphi\in S$. Two functions $\tilde{f}$ and $\tilde{g}$ in S' are said to be equal on a finite or semi-infinite *open* interval (a,b) if $\langle\tilde{f},\varphi\rangle=\langle\tilde{g},\varphi\rangle$ whenever $\varphi(x)=0$ for all $x\notin(a,b)$: one then writes $\tilde{f}=\tilde{g}$ on (a,b). $\tilde{f}\in S'$ is said to be regular on the finite or semi-infinite interval (a,b) if there exists an ordinary function f such that

$$\langle\tilde{f},\varphi\rangle=\int_a^b f(x)\varphi(x)\mathrm{d}x \tag{12.9}$$

whenever $\varphi(x)=0$ for all $x\notin(a,b)$. The delta functional $\tilde{\delta}$ is, for example, regular on $(-\infty,0)$ and on $(0,\infty)$, being equal to zero on these intervals. A functional $\tilde{f}$ in S' is said to be of bounded support if it is equal to zero on $(-\infty,a)$ and on (b,∞) for some values $-\infty<a<b<\infty$: the functional is then said to have its support on the *closed* interval $[a,b]$. Thus $\tilde{\delta}$ has its support on $[-1,1]$; also, less obviously, it has its support on $[0,1]$.

For every pair $\tilde{f}$ and $\tilde{g}$ of generalized functions in S' there exist generalized functions written as $\tilde{f}+\tilde{g}$ and $\tilde{f}-\tilde{g}$ which are called the *sum* and *difference*, respectively, defined so that for all $\varphi\in S$,

$$\langle\tilde{f}\pm\tilde{g},\varphi\rangle=\langle\tilde{f},\varphi\rangle\pm\langle\tilde{g},\varphi\rangle.$$

A sequence $\{\tilde{f}_n\}_{n=1}^{\infty}$ of generalized functions in S' is said to *converge in* S' if $\lim_{n\to\infty}\langle\tilde{f}_n,\varphi\rangle$ converges for each $\varphi\in S$, and very importantly it can be shown that there will then exist an $\tilde{f}\in S'$ such that

$$\lim_{n\to\infty}\langle\tilde{f}_n,\varphi\rangle=\langle\tilde{f},\varphi\rangle. \tag{12.10}$$

In this case we write $\lim_{n\to\infty}\tilde{f}_n=\tilde{f}$ (in S'), or we abbreviate still further and write $\tilde{f}_n\to\tilde{f}$ in S'. If $\tilde{f}_n\to\tilde{f}$ in S' and $\tilde{g}_n\to\tilde{g}$ in S' then it follows that $(\tilde{f}_n\pm\tilde{g}_n)\to(\tilde{f}\pm\tilde{g})$ in S'.

Addition and subtraction are commutative and associative so that always $\tilde{f}\pm\tilde{g}=\tilde{g}\pm\tilde{f}$ and $\tilde{f}+(\tilde{g}+\tilde{h})=(\tilde{f}+\tilde{g})+\tilde{h}$, for functionals in S'.

For each $\tilde{f}\in S'$ and each real or complex number a there exists a unique generalized function in S' written equivalently as $a\tilde{f}$ or as $\tilde{f}a$ with the property that $\langle a\tilde{f},\varphi\rangle=a\langle\tilde{f},\varphi\rangle$ for each $\varphi\in S$, and $a\tilde{f}$ is called the *product* of a and $\tilde{f}$. If $\tilde{f}_n\to\tilde{f}$, then also $a\tilde{f}_n\to a\tilde{f}$. Also, as might be expected, $a(\tilde{f}+\tilde{g})=a\tilde{f}+a\tilde{g}$, and $(a+b)\tilde{f}=a\tilde{f}+b\tilde{f}$, where a and b are numbers.

Given any $\tilde{f} \in S'$ there will exist a unique generalized function in S' written as $\tilde{f}'$ with the properties that for each $\varphi \in S$

$$\langle \tilde{f}', \varphi \rangle = \lim_{n\to\infty} \langle f'_n, \varphi \rangle \tag{12.11}$$

$$= -\langle \tilde{f}, \varphi' \rangle \tag{12.12}$$

where $\{f_n\}_{n=1}^{\infty}$ is any sequence of functions in S converging to $\tilde{f}$ in S', f'_n is the derivative of f_n, and φ' is the derivative of φ. Naturally $\tilde{f}'$ is called the *derivative* of $\tilde{f}$. It is easy enough to remember (12.11), and (12.12) becomes so when it is compared with the formula for integration by parts, which for functions ψ and φ in S has the form

$$\int_{-\infty}^{+\infty} \psi'\varphi = -\int_{-\infty}^{\infty} \psi\varphi'.$$

Clearly this result means that any $\tilde{f} \in S'$ can be differentiated any number of times. If $\tilde{f}$ and $\tilde{f}'$ are regular in S' and $\tilde{f} = f$, then it will follow that f is absolutely continuous and that $\tilde{f}' = f'$: generalized differentiation is thus consistent with ordinary differentiation in this specialized sense. If $\tilde{f}_n \to \tilde{f}$ in S', then it will follow that $\tilde{f}'_n \to \tilde{f}'$ in S'.

As an example of differentiation consider the signum function $f(x) = \frac{1}{2}\operatorname{sgn} x$. This defines a regular functional in S', say $\tilde{f}$. On differentiating we find that $\tilde{f}' = \tilde{\delta}$, the delta functional. One way of seeing this is by establishing first that

$$\lim_{n\to\infty} \pi^{-1} \arctan(nx) = \tilde{f} = f \quad (\text{in } S'),$$

so that on term by term differentiation

$$\lim_{n\to\infty} \frac{n}{\pi(n^2 + x^2)} = \tilde{f}' \quad (\text{in } S').$$

This sequence is well known as converging to the delta function, so that $\tilde{f}' = \tilde{\delta}$ as required. Alternatively we may utilize (12.12) as follows, ultimately integrating by parts:

$$\langle \tilde{f}', \varphi \rangle = -\langle \tilde{f}, \varphi' \rangle = -\frac{1}{2}\int_{-\infty}^{\infty} \operatorname{sgn} x\, \varphi'(x)\mathrm{d}x$$

$$= \frac{1}{2}\int_{-\infty}^{0} \varphi'(x)\mathrm{d}x - \frac{1}{2}\int_{0}^{\infty} \varphi'(x)\mathrm{d}x$$

$$= \varphi(0).$$

Since by definition $\langle \tilde{\delta}, \varphi \rangle = \varphi(0)$, we have $\tilde{f}' = \tilde{\delta}$ once again.

A product between two generalized functions $\tilde{f} \in S'$ and $\tilde{g} \in S'$ is not necessarily defined, but for arbitrary $\tilde{f} \in S'$ a product $\tilde{f}\tilde{g}$ is defined when $\tilde{g}$ is a special type of functional called a multiplier in S'. A functional $\tilde{g} \in S'$ is defined as a *multiplier in S′* if $\tilde{g}=g$ is regular in S' and if the ordinary product $g\varphi$ is in class S for each $\varphi\in S$. We give a direct characterization of multipliers in S' in theorem 14.2, but as examples the functions $\sin x$, $\cos x$, $(1+x^2)$ are multipliers in S', as is any function in S and any polynomial. The definition of the product $\tilde{f}\tilde{g}$ now relies on the following.

Theorem 12.2 Let $\tilde{f}$ be a functional in S' and $\tilde{g}$ be a multiplier in S': then there exists a unique functional in S', written as $\tilde{f}\tilde{g}$ or as $\tilde{g}\tilde{f}$, such that for all $\varphi\in S$,

$$\langle\tilde{f}\tilde{g},\varphi\rangle=\langle\tilde{f},g\varphi\rangle, \tag{12.13}$$

and

$$\langle\tilde{f}\tilde{g},\varphi\rangle=\lim_{n\to\infty}\int_{-\infty}^{\infty}f_n(x)g(x)\varphi(x)\mathrm{d}x \tag{12.14}$$

whenever $\{f_n\}$ is a sequence of ordinary functions converging in S' to $\tilde{f}$. □

The product $\tilde{f}\tilde{g}$ is also written as $\tilde{g}\tilde{f}$, $\tilde{f}g$ or $g\tilde{f}$, and in particular the product $\tilde{f}\varphi$ will exist for each $\tilde{f}\in S'$ and each $\varphi\in S$. When $\tilde{f}$ is regular, then the product $\tilde{f}\tilde{g}$ is equal to the ordinary product fg. As further examples, the product of $\tilde{\delta}$ with $\sin x$ is zero, whilst the product of $\tilde{\delta}$ with $\cos x$ is itself a delta functional. More examples appear in section 12.4. Equation (12.14) is a special case of the more general equation

$$\tilde{f}_n g\to\tilde{f}g \quad (\text{in } S'), \tag{12.15}$$

valid whenever $\tilde{f}\in S'$, g is a multiplier in S', and $\tilde{f}_n\to\tilde{f}$ in S'.

The ordinary and generalized derivatives of a multiplier in S' are equal, and the derivative will itself be a multiplier in S'. It can also be shown that for any $\tilde{f}\in S'$ and any multiplier g in S',

$$(\tilde{f}g)'=\tilde{f}'g+\tilde{f}g'. \tag{12.16}$$

A generalized function can be *shifted* as follows. Given $\tilde{f}\in S'$, and some real number a, there exists a unique generalized function, say $\tilde{f}_a$, with the property that, for each $\varphi\in S$,

$$\begin{aligned}\langle\tilde{f}_a,\varphi\rangle&=\lim_{n\to\infty}\int_{-\infty}^{\infty}f_n(x-a)\varphi(x)\mathrm{d}x\\&=\langle\tilde{f},\varphi_a\rangle,\end{aligned} \tag{12.17}$$

where $\{f_n\}_{n=1}^{\infty}$ is any sequence of ordinary functions such that $f_n\to\tilde{f}$ in S', and φ_a is defined by $\varphi_a(x)=\varphi(x+a)$. The generalized function $\tilde{f}_a$ behaves like a version of $\tilde{f}$ that is shifted a distance a along the x-axis, and it is sometimes convenient to write $\tilde{f}_a$ as $\tilde{f}(x-a)$. For example, the shifted delta function, $\tilde{\delta}_a$ or $\tilde{\delta}(x-a)$, has the property that for each $\varphi\in S$, $\langle\tilde{\delta}_a,\varphi\rangle=\varphi(a)$.

It can be shown that for each $\tilde{f} \in S'$ there exists a unique functional in S' written as $\tilde{f}_{\mathrm{T}}$ such that for each $\varphi \in S$,

$$\begin{aligned}\langle \tilde{f}_{\mathrm{T}}, \varphi \rangle &= \lim_{n\to\infty} \int_{-\infty}^{\infty} f_n(-x)\varphi(x)\mathrm{d}x \\ &= \langle \tilde{f}, \varphi_{\mathrm{T}} \rangle, \end{aligned} \tag{12.18}$$

where $\{f_n\}$ is any sequence of ordinary functions converging to $\tilde{f}$ in S', and where φ_{T} is the transpose of φ defined by $\varphi_{\mathrm{T}}(x) = \varphi(-x)$. $\tilde{f}_{\mathrm{T}}$ is called the *transpose* of $\tilde{f}$.

A generalized function can be *scaled* as follows. Given $\tilde{f} \in S'$, and some non-zero real number a, there exists a unique generalized function, say $\tilde{f}_a$, with the property that for each $\varphi \in S$

$$\begin{aligned}\langle \tilde{f}_a, \varphi \rangle &= \lim_{n\to\infty} \int_{-\infty}^{+\infty} f_n(ax)\varphi(x)\mathrm{d}x \\ &= \langle \tilde{f}, \varphi_a \rangle \end{aligned} \tag{12.19}$$

where $\{f_n\}_{n=1}^{\infty}$ is any sequence such that $f_n \to \tilde{f}$ in S', and φ_a is defined by $\varphi_a(x) = |a|^{-1}\varphi(ax)$. The generalized function $\tilde{f}_a$ can conveniently be written as $\tilde{f}(ax)$.

For each $\tilde{f} \in S'$ there exists a unique generalized function in S', written $\tilde{f}^*$, with the properties that for each $\varphi \in S$

$$\begin{aligned}\langle \tilde{f}^*, \varphi \rangle &= \lim_{n\to\infty} \int_{-\infty}^{+\infty} f_n^*(x)\varphi(x)\mathrm{d}x \\ &= \{\langle \tilde{f}, \varphi^* \rangle\}^*, \end{aligned} \tag{12.20}$$

where $\{f_n\}_{n=1}^{\infty}$ is any sequence of functions such that $f_n \to \tilde{f}$, in S' and the asterisk means complex conjugation. Naturally $\tilde{f}^*$ is called the *complex conjugate* of $\tilde{f}$.

If $\tilde{f}$ and $\tilde{g}$ are functionals in S', and if $\tilde{f} = \tilde{g}'$ then $\tilde{g}$ is said to be a *primitive* or *indefinite integral* of $\tilde{f}$. It can be shown that each $\tilde{f} \in S'$ possesses an infinite set of primitives which are unique apart from an additive constant. For example, the function $\frac{1}{2}\operatorname{sgn} x$ and the Heaviside step function $H(x) = \frac{1}{2}(\operatorname{sgn} x + 1)$ are each a primitive of the delta functional. If $\tilde{f}$ is regular then this definition is consistent with the ordinary definition in section 3.6. The following theorem provides a method of specifying the primitive of a functional in S'.

Theorem 12.3 Let $\tilde{f}$ be a functional in S' and let ψ be some function in S such that $\int \psi = 1$: then (i) for each $\varphi \in S$ the function χ defined at all x by

$$\chi(x) = \int_{-\infty}^{x} \left\{ \varphi(u) - \left[\int \varphi \right] \psi(u) \right\} \mathrm{d}u$$

will be in class S, (ii) the functional $\tilde{g}$ defined so that for each $\varphi \in S$ and some fixed ψ,

$$\langle \tilde{g}, \varphi \rangle = \langle \tilde{f}, \chi \rangle$$

is a primitive of $\tilde{f}$, and (iii) different choices for ψ lead to primitive differing by a constant. □

Clearly a functional in S' can be integrated in the above sense any number of times. Some further results relating to repeated integration are given in section 14.1.

12.4 Examples of generalized functions

We now give examples of some of the more common generalized functions. In each case it is necessary to specify the value of $\langle \tilde{f}, \varphi \rangle$ for all $\varphi \in S$, and this can be done (i) by a direct rule, or (ii) by specifying $\tilde{f}$ as the nth derivative of some ordinary function (see theorem 14.5), or (iii) by means of a sequence of ordinary functions which converge to $\tilde{f}$ in S', or (iv) by a sequence of functionals converging to $\tilde{f}$ in S'. Method (iii) often gives a good insight into the nature of $\tilde{f}$, whilst the first two methods are often neater. In view of theorem 9.1 every ordinary function in class K corresponds to a regular functional in S'.

The delta functional may be defined from any of the following equations, since it can be shown that they all define the same generalized function:

$$\langle \tilde{\delta}, \varphi \rangle = \varphi(0) \quad \text{for each } \varphi \in S$$

$$\tilde{\delta} = (H(x))' = \tfrac{1}{2}(\operatorname{sgn} x)'.$$

Here the Heaviside function H, and the signum function sgn, are each in K and thus define functionals in S'. If f is a function in $L(-\infty, \infty)$ and $\int f = 1$, then other representations of $\tilde{\delta}$ follow from

$$\tilde{\delta}(x) = \lim_{a \to \infty} af(ax) \quad (\text{in } S').$$

For example, as $a \to \infty$ so each of the following converges in S' to $\tilde{\delta}(x)$: $a\exp(-\pi a^2 x^2)$, $(a/2)\exp(-a|x|)$, $aH(x)\exp(-ax)$, and $(a/2)\operatorname{rect}(ax)$, where $\operatorname{rect}(x)$ is equal to unity on $(-1, 1)$ and zero elsewhere. Also

$$\tilde{\delta}(x) = \lim_{\varepsilon \to 0^+} \frac{\varepsilon}{\pi(x^2 + \varepsilon^2)} \quad (\text{in } S').$$

It is not necessary that a sequence must consist of functions in L, and for instance the following can be established:

$$\tilde{\delta}(x) = \lim_{a \to \infty} \frac{\sin ax}{\pi x} \quad (\text{in } S')$$

and

$$\tilde{\delta}(x) = \lim_{a \to \infty} a\pi^{-1/2} \exp[\mathrm{i}(a^2x^2 - \pi/4)] \quad (\text{in } S').$$

The product of $\tilde{\delta}$ with any multiplier in S' exists, and in particular $x\tilde{\delta}=0$. If $g(x)$ is a multiplier, then the product $\tilde{g}\tilde{\delta}$ is equal to $g(0)\tilde{\delta}$, and $\tilde{g}(x)\tilde{\delta}(x-a)$ is equal to $g(a)\tilde{\delta}(x-a)$.

Since each generalized function can be differentiated any number of times, we can form the differentials of $\tilde{\delta}$, and these are written δ', δ'', δ''', ..., $\delta^{(n)}$, ... These are useful in representing dipoles, quadrupoles, octupoles and 2^n-poles in physical situations. One can visualize $\tilde{\delta}'$ as having a graph which is the limiting case of a tall narrow hump just to the left of the origin with a similar negative hump just to the right. More formally, it can be shown that:

$$\delta^{(n)}(x)=(H(x))^{(n+1)}=\tfrac{1}{2}(\operatorname{sgn} x)^{(n+1)}$$

$$\langle \tilde{\delta}^{(n)}, \varphi\rangle=(-1)^n\varphi^{(n)}(0)$$

$$\tilde{\delta}'(x)=\lim_{\varepsilon\to 0^+}\left(\frac{-2\varepsilon x}{x^2+\varepsilon^2}\right)$$

$$=\lim_{a\to\infty}\left(\frac{ax\cos ax-\sin ax}{\pi x^2}\right)$$

$$\tilde{\delta}^{(n)}(x)=\lim_{\varepsilon\to 0^+}\frac{\mathrm{d}^n}{\mathrm{d}x^n}\frac{\varepsilon}{\pi(x^2+\varepsilon^2)},$$

where the limits are in S'. Since $\tilde{\delta}'$ may be imagined as a double hump at the origin, it might be thought that the product $x\tilde{\delta}'$ would be zero. This is not so, and in fact $x\tilde{\delta}'=-\tilde{\delta}$, and, for real a,

$$x\tilde{\delta}'(x-a)=a\tilde{\delta}'(x-a)-\tilde{\delta}(x-a).$$

Indeed if $\tilde{g}=g$ is a multiplier in S' then we have the relation

$$\tilde{g}\tilde{\delta}^{(n)}=\sum_{p=0}^{n}A_p g^{(p)}(0)\tilde{\delta}^{(n-p)} \tag{12.21}$$

where the numbers A_p are given by

$$A_p=\frac{(-1)^p n!}{p!(n-p)!}. \tag{12.22}$$

For a further result along these lines see (13.27).

In section 12.3 we have given meaning to the statement that two functionals in S' are equal on an open interval, and it can be shown that in this sense each of the $\tilde{\delta}^{(n)}$, $n=0, 1, 2, \ldots$, is equal to zero on $(-\infty, 0)$ and on $(0, \infty)$. Indeed it can be shown that if $\tilde{f}$ is some generalized function equal to zero on $(-\infty, 0)$ and $(0, \infty)$ then $\tilde{f}$ must necessarily be equal to a finite sum of the type

$$\tilde{f}=a_0\tilde{\delta}+a_1\tilde{\delta}'+\ldots+a_n\tilde{\delta}^{(n)},$$

where $a_0, a_1, \ldots$ are real or complex numbers, and where the value of n

depends on $\bar{f}$. Likewise if two generalized functions are equal on $(-\infty, 0)$ and on $(0, \infty)$, then their difference must be of the above form.

The effect on a delta function of scaling the x-variable is not quite obvious, and we therefore quote the following result, which is valid for any non-zero real number, a and for $n=0, 1, 2, \ldots$:

$$\delta^{(n)}(ax)=|a|^{-1}a^{-n}\delta^{(n)}(x).$$

The function $f(x)=x^n$, $n=0, 1, 2, 3, \ldots$, is in class K and thus defines a regular functional in S' which is written as x^n. For $m=1, 2, 3, \ldots$ the symbol x^{-m} is used to represent singular generalized functions in S' as follows, by means of the generalized derivatives of $\ln|x|$:

$$x^{-m}=A_m(\ln|x|)^{(m)}, \tag{12.23a}$$

where

$$A_m=\frac{(-1)^{m-1}}{(m-1)!}. \tag{12.23b}$$

This definition is in order because the function $\ln|x|$ is in class K. The symbol x^{-m} is a natural choice for these generalized functions because (12.23) is valid as an ordinary equation when $x\neq 0$ and when the generalized derivative is replaced by an ordinary derivative. Moreover, if $\varphi\in S$ is zero on an interval $(-\varepsilon, \varepsilon)$, $\varepsilon>0$, then

$$\langle x^{-m}, \varphi\rangle=\int_{-\infty}^{+\infty} x^{-m}\varphi(x)\mathrm{d}x.$$

The function x^n $(n=0, 1, 2, \ldots)$ is a multiplier in S' and so the generalized product $x^n x^{-m}$ is defined, and it can be shown that this product is indeed equal to the functional x^{-m+n}. The following relations between functionals are also valid:

$$(x^{-m})'=-mx^{-m-1}$$

$$(ax)^{-m}=a^{-m}x^{-m} \qquad (a>0).$$

The functional x^{-1} is the most commonly met member of the set just defined, and the following relations give an insight into its nature. For each $\varphi\in S$:

$$\langle x^{-1}, \varphi\rangle=\lim_{\varepsilon\to 0^+}\int_{|x|>\varepsilon}\frac{\varphi(x)}{x}\mathrm{d}x \tag{12.24}$$

$$=\int_0^{\infty}\frac{\varphi(x)-\varphi(-x)}{x}\mathrm{d}x \tag{12.25}$$

$$=-\int_{-\infty}^{\infty}\ln|x|\varphi'(x)\mathrm{d}x. \tag{12.26}$$

$$x^{-1}=\lim_{\varepsilon\to 0^+}\frac{x}{x^2+\varepsilon^2} \quad (\text{in } S'). \tag{12.27}$$

The limit in (12.24) represents a Cauchy principal value of the integral at $x=0$, and for this reason the symbol P/x is sometimes used instead of x^{-1} for this generalized function.

The functional x^{-2} cannot be described by a Cauchy principal value, but the following relations are illuminating. For each $\varphi \in S$:

$$\langle x^{-2}, \varphi \rangle = \int_0^\infty \frac{\varphi(x)+\varphi(-x)-2\varphi(0)}{x^2}\,\mathrm{d}x. \tag{12.28}$$

$$x^{-2} = \lim_{\varepsilon \to 0^+} \frac{x^2-\varepsilon^2}{(x^2+\varepsilon^2)^2} \quad (\text{in } S'). \tag{12.29}$$

The integral in (12.28) converges absolutely as an ordinary Lebesgue integral. Note that the sequence in (12.29) will no longer converge in S' if the numerator, $(x^2-\varepsilon^2)$, is replaced by x^2. More generally for $m=1,2,3,\ldots$ one has

$$x^{-m} = \lim_{\varepsilon \to 0^+} A_m \frac{\mathrm{d}^m}{\mathrm{d}x^m} \frac{\ln(x^2+\varepsilon^2)}{2}, \tag{12.30}$$

where A_m is given in (12.23*b*).

We now consider non-integer negative powers of x. For $0<\beta<1$ the functions $|x|^{-\beta}$, $|x|^{-\beta}\operatorname{sgn} x$, $|x|^{-\beta}H(x)$, and $|x|^{-\beta}H(-x)$ are in class K and so define regular generalized functions in S' whose derivatives may be used to define other generalized functions in S'. For $m=1, 2, 3, \ldots$, $|x|^{-m-\beta}$, $|x|^{-m-\beta}\operatorname{sgn} x$, and $|x|^{-m-\beta}H(\pm x)$ are used to represent functionals in S' defined by

$$|x|^{-\beta-m} = B(m,\beta)[|x|^{-\beta}(\operatorname{sgn} x)^m]^{(m)} \tag{12.31}$$

$$|x|^{-\beta-m}\operatorname{sgn} x = B(m,\beta)[|x|^{-\beta}(\operatorname{sgn} x)^{m+1}]^{(m)} \tag{12.32}$$

$$|x|^{-\beta-m}H(\pm x) = \tfrac{1}{2}[|x|^{-\beta-m} \pm |x|^{-\beta-m}\operatorname{sgn} x], \tag{12.33}$$

where

$$B(m,\beta) = \frac{(-1)^m}{\beta(\beta+1)(\beta+2)\ldots(\beta+m-1)}.$$

The value of $B(m,\beta)$ is just that required to make (12.31) and (12.32) valid as ordinary differentials when $x \neq 0$, so the symbols are natural ones.

These functionals behave just as one might hope, in certain simple contexts. For instance, for each $\varphi \in S$ that is zero on $(-\varepsilon, \varepsilon)$, for some $\varepsilon>0$, we find that

$$\langle |x|^{-m-\beta}K, \varphi \rangle = \int_{-\infty}^{\infty} |x|^{-m-\beta}K\varphi(x)\,\mathrm{d}x,$$

where K may stand in place of 1, $\operatorname{sgn} x$, $H(x)$ or $H(-x)$. The following generalized equations are formally the same as for ordinary products, for

$m=1,2,3,\ldots$.

$$x^n|x|^{-m-\beta}=\begin{cases}|x|^{-m-\beta+n}, & n=0,2,4,6,\ldots\\ |x|^{-m-\beta+n}\operatorname{sgn}x, & n=1,3,5,7,\ldots\end{cases}$$

$$x^n|x|^{-m-\beta}\operatorname{sgn}x=\begin{cases}|x|^{-m-\beta+n}\operatorname{sgn}x, & n=0,2,4,6,\ldots\\ |x|^{-m-\beta+n}, & n=1,3,5,7,\ldots\end{cases}$$

Scaling the x variable by a factor $a>0$ also has the same effect as with ordinary functions, and

$$|ax|^{-m-\beta}=a^{-m-\beta}|x|^{-m-\beta}.$$

$$|ax|^{-m-\beta}\operatorname{sgn}x=a^{-m-\beta}|x|^{-m-\beta}\operatorname{sgn}x.$$

One way of expressing these functionals as a limit in S' of a sequence of integrable functions is as follows:

$$|x|^{-m-\beta}=\lim_{\varepsilon\to 0^+}\frac{\cos[(m+\beta)\arctan(x/\varepsilon)]}{(x^2+\varepsilon^2)^{(m+\beta)/2}\cos[(m+\beta)\pi/2]}$$

$$|x|^{-m-\beta}\operatorname{sgn}x=\lim_{\varepsilon\to 0^+}\frac{\sin[(m+\beta)\arctan(x/\varepsilon)]}{(x^2+\varepsilon^2)^{(m+\beta)/2}\sin[(m+\beta)\pi/2]},$$

where, as usual, we choose $-\pi/2<\arctan<\pi/2$.

The symbols $(x+\mathrm{i}0)^\alpha$ and $(x-\mathrm{i}0)^\alpha$ are used to represent functionals in S' for each complex value of α. These are based on the limits as $\varepsilon\to 0^+$ of the functions $(x+\mathrm{i}\varepsilon)^\alpha$ and $(x-\mathrm{i}\varepsilon)^\alpha$, each of which is in class K for $\varepsilon>0$. To remove the multi-valuedness of these quantities, the functionals are defined through the following limit in S', for each complex α,

$$(x\pm\mathrm{i}0)^\alpha=\lim_{\varepsilon\to 0^+}\exp\{\alpha[\ln|x\pm\mathrm{i}\varepsilon|+\mathrm{i}\arg(x\pm\mathrm{i}\varepsilon)]\},$$

where $-\pi<\arg<\pi$. When α is real this becomes

$$(x\pm\mathrm{i}0)^\alpha=\lim_{\varepsilon\to 0^+}(x^2+\varepsilon^2)^{\alpha/2}\exp\{\pm\mathrm{i}\alpha[(\pi/2)-\arctan(x/\varepsilon)]\}, \quad (12.34)$$

where $-\pi/2<\arctan<\pi/2$. For simplicity we will restrict further discussion to the case of real α.

For $\alpha>-1$ these functionals are regular, and then

$$(x\pm\mathrm{i}0)^\alpha=\begin{cases}|x|^\alpha, & x>0\\ |x|^\alpha\exp(\pm\mathrm{i}\alpha\pi), & x<0.\end{cases}$$

For $\alpha\leqslant -1$ the functionals are singular. The most commonly met case is with $\alpha=-1$, and there is a relationship with previously defined functionals as follows:

$$(x\pm\mathrm{i}0)^{-1}=x^{-1}\mp\mathrm{i}\pi\delta(x).$$

This is a special case of the relation, for $m=1,2,3,\ldots$

$$(x\pm\mathrm{i}0)^{-m}=x^{-m}\mp\mathrm{i}A_m\pi\delta^{(m-1)}(x),$$

where A_m has the value in (12.23*b*). Another link with previously defined functionals is as follows:

$$(x \pm i0)^\alpha = |x|^\alpha H(x) + \exp(\pm i\alpha\pi)|x|^\alpha H(-x),$$

this being valid for all real α except $\alpha = -1, -2, -3, \ldots$. Finally, generalized differentiation gives an answer analogous to that for ordinary differentiation, for all real α,

$$[(x \pm i0)^\alpha]' = \alpha(x \pm i0)^{\alpha-1}.$$

We now consider some singular functionals in S' written as $x^{-m}\operatorname{sgn} x$, for $m = 1, 2, 3, \ldots$. Authors differ somewhat in the functional represented by this symbol, but in all cases the definition can be written as

$$x^{-m}\operatorname{sgn} x = A_m[(\ln|x|\operatorname{sgn} x)^{(m)} + B_m\delta^{(m-1)}(x)]. \tag{12.35}$$

The coefficient A_m is as in (12.23*b*), whilst the numbers B_m are chosen differently by different authors. One choice is to put $B_m = 0$ for each m: this makes the definition rather analogous to that given in (12.23) for x^{-m}, and it also leads to the attractive relation

$$(x^{-m}\operatorname{sgn} x)' = -m(x^{-m-1}\operatorname{sgn} x). \tag{12.36}$$

However, the choice $B_m = 0$ leads to the result that the generalized product $x(x^{-m}\operatorname{sgn} x)$ is *not* equal to $x^{-m+1}\operatorname{sgn} x$ for $m = 2, 3, 4, \ldots$ (though it is for $m = -1$).

An alternative choice is to put $B_1 = 0$, $B_2 = 2$ and

$$B_m = 2\sum_{k=1}^{m-1}\frac{1}{k} \qquad (m = 3, 4, \ldots). \tag{12.37}$$

This choice has the merit that for $n = 0, 1, 2, \ldots$ and $m = 1, 2, 3, \ldots$

$$x^n(x^{-m}\operatorname{sgn} x) = x^{-m+n}\operatorname{sgn} x, \tag{12.38}$$

and this convention is used for instance by Gel'fand *et al.* (1964–9) and by Zemanian (1965). However, the differentiation rule is now more complicated than that in (12.36). Formulae valid for any convention are as follows:

$$(x^{-m}\operatorname{sgn} x)' = -mx^{-m-1}\operatorname{sgn} x - A_m[B_{m+1} - B_m]\delta^{(m)}(x)$$

$$x(x^{-m}\operatorname{sgn} x) = x^{-m+1}\operatorname{sgn} x + A_m[2 - (m-1)(B_{m+1} - B_m)]\delta^{(m)}(x),$$

with A_m given by (12.23*b*).

Another convention, adopted for instance by Lighthill (1959) and by Jones (1966), is to leave the choice of B_m, at each m, arbitrary: this means that at each m the symbol $x^{-m}\operatorname{sgn} x$ represents a family of functionals differing from each other by a multiple of $\delta^{(m-1)}(x)$. There is an analogy with the way in which an indefinite integral is defined apart from an additive constant and thus represents a family of functions. Under this convention (12.36) and (12.38) both hold true, provided they are interpreted as meaning that at each $m=1,2,3,\ldots$ the class of functionals on the left side is the same as the class of functionals on the right. This convention has the further advantage that on scaling the x-variable we have, for $a>0$,

$$(ax)^{-m}\operatorname{sgn} x=a^{-m}x^{-m}\operatorname{sgn} x, \tag{12.39}$$

a result which does not hold for either of the other conventions given above; once again (12.39) represents a relation between classes of functionals.

The most commonly met functional of this type is $x^{-1}\operatorname{sgn} x$. Adopting the convention $B_1=0$ we have for each $\varphi\in S$,

$$\begin{aligned}\langle x^{-1}\operatorname{sgn} x,\varphi\rangle &= \lim_{\varepsilon\to 0^+}\left[2\varphi(0)\ln\varepsilon+\int_{|x|>\varepsilon}\frac{\varphi(x)}{|x|}\,\mathrm{d}x\right]\\ &=\int_{-1}^{+1}\frac{\varphi(x)-\varphi(0)}{|x|}\,\mathrm{d}x+\int_{|x|>1}\frac{\varphi(x)}{|x|}\,\mathrm{d}x,\end{aligned}$$

and

$$\begin{aligned}x^{-1}\operatorname{sgn} x&=(\ln|x|\operatorname{sgn} x)'\\ &=\lim_{\varepsilon\to 0^+}\left[\frac{2x\arctan(x/\varepsilon)+\varepsilon\ln(x^2+\varepsilon^2)}{\pi(x^2+\varepsilon^2)}\right]\quad(\text{in } S').\end{aligned}$$

The singular generalized functions $x^{-m}H(x)$ and $x^{-m}H(-x)$ are defined, for $m=1,2,3,\ldots$, by

$$x^{-m}H(\pm x)=\tfrac{1}{2}[x^{-m}\pm x^{-m}\operatorname{sgn} x].$$

Since this definition relies on the definition of $x^{-m}\operatorname{sgn} x$ it is clear that the various conventions lead to different meanings for $x^{-m}H(\pm x)$.

A contrast between the functionals x^{-m} and the functionals $x^{-m}\operatorname{sgn} x$ appears in the following sequences in S', where λ is positive and not equal to an integer:

$$\lim_{\lambda\to m}|x|^{-\lambda}=\begin{cases}x^{-m}, & m=2,4,6,\ldots\\ \text{divergent in } S', & m=1,3,5,\ldots\end{cases}$$

$$\lim_{\lambda\to m}|x|^{-\lambda}\operatorname{sgn} x=\begin{cases}x^{-m}, & m=1,3,5,\ldots\\ \text{divergent in } S', & m=2,4,6,\ldots.\end{cases}$$

We see that divergent behaviour occurs for those values of m at which one might have expected the limit functional $x^{-m}\operatorname{sgn} x$. Although the symbol $|x|^{-1}$ is often used to represent the functional $x^{-1}\operatorname{sgn} x$, its use must not be taken to imply that $\lim_{\varepsilon\to 0_+}|x|^{-1\pm\varepsilon}=|x|^{-1}$ in S'. We will avoid the use of the symbols $|x|^{-1}$, $|x|^{-3}$, $|x|^{-5}$, $|x|^{-2}\operatorname{sgn} x$, $|x|^{-4}\operatorname{sgn} x$, etc.

Having dealt with inverse powers of x, it is possible to interpret all rational functions as generalized functions by the use of partial fractions. Consider, for example, the ordinary function $f(x)=2(x^2-1)^{-1}$, $(x\neq 1)$.

This is not in class K and does not represent a regular functional in S'. However, on writing

$$\frac{2}{x^2-1}=\frac{1}{x-1}-\frac{1}{x+1}$$

and recognizing that the right hand side can be interpreted as a difference between shifted versions of the functional x^{-1} we achieve a definition for the functional $2(x^2-1)^{-1}$.

Further rules for associating functionals with ordinary functions that are not locally integrable are based on Hadamard's 'finite part' of a divergent integral. These rules are described for instance in Zemanian (1965). A functional in S' is then associated, for example, with the ordinary function $(\sin x)^{-1}$.

13

Fourier transformation of generalized functions – I

13.1 Definition of the transform

The Fourier transform of a functional in S' is easy to define on account of the following theorem.

Theorem 13.1 Let $\tilde{f}$ be a functional in S': then there exists a functional, say $\tilde{F}$, also in S' such that for each $\Phi \in S$,

$$\langle \tilde{F}, \Phi \rangle = \langle \tilde{f}, \varphi_{\mathrm{T}} \rangle \tag{13.1}$$

and

$$\langle \tilde{F}, \Phi^* \rangle = \langle \tilde{f}, \varphi^* \rangle \tag{13.2}$$

where φ is the inverse transform of Φ, (7.2), and φ_{T} is the transpose of φ. □

Equations (13.1) and (13.2) are clearly reminiscent of the Parseval equations for ordinary Fourier pairs and accordingly we *define* $\tilde{F}$ to be the Fourier transform in S' of $\tilde{f}$ if (13.1) and/or (13.2) is valid for every pair $\varphi \leftrightarrow \Phi$ in S; we write $\tilde{f} \leftrightarrow \tilde{F}$ in S', and say that $\tilde{f}$ and $\tilde{F}$ form a Fourier pair in S', $\tilde{f}$ being the inverse transform of $\tilde{F}$. The definition is unique in the sense that if $\tilde{f} \leftrightarrow \tilde{F}$ and $\tilde{g} \leftrightarrow \tilde{F}$ then $\tilde{f} = \tilde{g}$, and if $\tilde{f} \leftrightarrow \tilde{F}$ and $\tilde{f} \leftrightarrow \tilde{G}$ then $\tilde{F} = \tilde{G}$. The definition is also consistent with our previous definition of Fourier pairs in K, in that if $f \leftrightarrow F$ in K then f and F define a regular Fourier pair $\tilde{f} \leftrightarrow \tilde{F}$ in S' and conversely.

The convergence of a sequence of functions in S' is accompanied by convergence of the transforms, a result leading to great simplicity.

Theorem 13.2 Let $\{\tilde{f}_n\}_{n=1}^{\infty}$ be a sequence of functionals in S', and suppose that $\lim_{n\to\infty} \tilde{f}_n = \tilde{f}$ in S': then it will follow that $\lim_{n\to\infty} \tilde{F}_n = \tilde{F}$, where $\tilde{f}_n \leftrightarrow \tilde{F}_n$ and $\tilde{f} \leftrightarrow \tilde{F}$ in S', and analogously if n is replaced by a continuous parameter. □

This theorem can be of great use in finding the transform of a functional in S', since if we choose a sequence of functions f_n each of which is in L then the transforms F_n can be found by ordinary integration, and $\tilde{F}$ is then determined as the limit in S' of the F_n. For example, if $f_\varepsilon(x)=\exp(-2\pi\varepsilon|x|)$, $\varepsilon>0$, we have $F_\varepsilon(y)=(\varepsilon/\pi)(y^2+\varepsilon^2)^{-1}$, whence when $\varepsilon\to 0^+$ we find the pair $1\leftrightarrow\tilde{\delta}$ in S'. We can confirm the validity of this result as follows by appealing directly to the definition based on theorem 13.1. For each pair $\varphi\leftrightarrow\Phi$ in S,

$$\langle 1,\varphi\rangle=\int\varphi=\Phi(0)=\Phi_{\mathrm{T}}(0)=\langle\tilde{\delta},\Phi_{\mathrm{T}}\rangle,$$

so that $1\leftrightarrow\tilde{\delta}$. In some treatments the definition of the Fourier transform of a functional in S' is based on sequences of good functions: this is possible because every functional in S' can be expressed as the limit, in S', of a sequence of good functions, and the transform is then defined as the limit in S' of the transforms of the good functions.

13.2 Simple properties of the transform

Many results valid for ordinary transforms have their counterparts with functionals in S'. The transformation is *linear*, in that for any complex α and β, and any pairs $\tilde{f}\leftrightarrow\tilde{F}$ and $\tilde{g}\leftrightarrow\tilde{G}$ in S' it follows that

$$(\alpha\tilde{f}+\beta\tilde{g})\leftrightarrow(\alpha\tilde{F}+\beta\tilde{G}).$$

The *shift theorem* states that if $\tilde{f}\leftrightarrow\tilde{F}$ in S', and if a is a real number then

$$\tilde{f}(x-a)\leftrightarrow\exp(-2\pi i a y)\tilde{F}(y)$$
$$\exp(+2\pi i a x)\tilde{f}(x)\leftrightarrow\tilde{F}(y-a).$$

The *scaling* theorem states that if $\tilde{f}\leftrightarrow\tilde{F}$ in S', and if b is a non-zero real number, then

$$|b|\tilde{f}(bx)\leftrightarrow\tilde{F}(y/b).$$

The operation of *transposition*, $\tilde{f}_{\mathrm{T}}(x)=\tilde{f}(-x)$, and complex *conjugation* behave as follows on Fourier transformation, just as in the case of ordinary Fourier pairs. Given $\tilde{f}\leftrightarrow\tilde{F}$, in S', then $\tilde{f}_{\mathrm{T}}\leftrightarrow\tilde{F}_{\mathrm{T}}$, $\tilde{f}^*\leftrightarrow\tilde{F}^*_{\mathrm{T}}$, $\tilde{F}\leftrightarrow\tilde{f}_{\mathrm{T}}$, and $\tilde{F}_{\mathrm{T}}\leftrightarrow\tilde{f}$.

Differentiation is simpler to deal with than is the case with ordinary functions since every functional is differentiable.

Theorem 13.3 Suppose $\tilde{f}\leftrightarrow\tilde{F}$ in S' and that $\tilde{f}^{(n)}$ and $\tilde{F}^{(n)}$ are the nth differentials of $\tilde{f}$ and $\tilde{F}$, $n=0,1,2,\ldots$: then it follows that for each n:

$$\tilde{f}^{(n)}(x)\leftrightarrow(2\pi i y)^n\tilde{F}(y)$$
$$(-2\pi i x)^n\tilde{f}(x)\leftrightarrow\tilde{F}^{(n)}(y).\quad\square$$

As an example of the use of this theorem, starting from the pair $1\leftrightarrow\tilde{\delta}$, if we repeatedly differentiate the right hand member we obtain, for $n=0, 1, 2, \ldots,$

$$(-2\pi i x)^n \leftrightarrow \tilde{\delta}^{(n)}. \tag{13.3}$$

As a converse to the differentiation theorem one might wonder whether multiplying a functional by $(-2\pi i x)^{-1}$ will have the effect of integrating the transform. However, we have so far only given meaning to the product $x^{-1}\tilde{f}(x)$ when $\tilde{f}$ is a multiplier, section 12.3, and in this case we have the following.

Theorem 13.4 Let $\tilde{f}$ be a multiplier in S' with transform $\tilde{F}$: then $(-2\pi i)^{-1}x^{-1}\tilde{f}(x)$ and $(-2\pi i)^{-1}(x \pm i0)^{-1}\tilde{f}(x)$ are functionals in S' whose Fourier transforms are in each case equal to an indefinite integral of $\tilde{F}$. □

As an example we may choose a good function φ as the multiplier, and it then happens that

$$(-2\pi i)^{-1}x^{-1}\varphi(x) \leftrightarrow \int_{-\infty}^{y} \Phi(u)\mathrm{d}u - \varphi(0)/2$$

and

$$(-2\pi i)^{-1}(x+i0)^{-1}\varphi(x) \leftrightarrow \int_{-\infty}^{y} \Phi(u)\mathrm{d}u,$$

where Φ is the transform of φ.

13.3 Examples of Fourier transforms

Extensive tabulations of transforms of functionals are to be found, for instance, in Gel'fand *et al.* (1964–9) and Jones (1966), and the table below is intended as a short illustrative list of Fourier pairs in S'. Although as these authors show an extension to complex powers of x is possible, we will restrict ourselves to real powers, and in the table we assume $n=0, 1, 2, 3, \ldots$; $m=1, 2, 3, \ldots$; $0<\alpha<\infty$; $\lambda>0$ and $\lambda \neq 1, 2, 3, \ldots$; whilst a is any real number. The gamma function $\Gamma(\alpha)$ is tabulated for instance in Abramowitz and Stegun (1966) and $\Gamma(m)=(m-1)!$, whilst the function $\psi(m)$ is defined subsequently in (13.14).

In Table 13.1 we have arranged that the functionals $\tilde{f}$ are regular and the functionals $\tilde{F}$ are singular, but the roles may be reversed on noting that $\tilde{f} \leftrightarrow \tilde{F}$ implies $\tilde{F} \leftrightarrow \tilde{f}_T$, where $\tilde{f}_T$ is the transpose of $\tilde{f}$. There are many ways of establishing the above transforms, and we list now just one method in each case. Although the integrals referred to are not entirely well known, they are all to be found in Gradshtein and Ryzhik (1965). Transform (13.5) has already been established in sections 13.1 and 13.2, (13.3).

Transform (13.6) is established from the following pair, in which the left

Table 13.1. *Table of Fourier pairs in S′*

$\tilde{f}(x)$	$\tilde{F}(y)$	
$f(x)$, absolutely integrable	$\int_{-\infty}^{+\infty} f(x)\exp(-2\pi i x y)\mathrm{d}x$	(13.4)
x^n	$(-2\pi i)^{-n}\delta^{(n)}(y)$	(13.5)
$x^{m-1}\operatorname{sgn} x$	$\dfrac{2(m-1)!}{(2\pi i)^m}\, y^{-m}$	(13.6)
$x^{m-1}[\ln\|2\pi x\|-\psi(m)]$	$\dfrac{-\pi i(m-1)!}{(2\pi i)^m}\, y^{-m}\operatorname{sgn} y$	(13.7)
$x^{\alpha-1}H(\mp x)$	$\dfrac{\Gamma(\alpha)\exp(\pm i\alpha\pi/2)}{(2\pi)^\alpha}(y\pm i0)^{-\alpha}$	(13.8)
$\|x\|^{\lambda-1}$	$\dfrac{2\Gamma(\lambda)\cos(\lambda\pi/2)}{(2\pi)^\lambda}\|y\|^{-\lambda}$	(13.9)
$\|x\|^{\lambda-1}\operatorname{sgn} x$	$\dfrac{-2i\Gamma(\lambda)\sin(\lambda\pi/2)}{(2\pi)^\lambda}\|y\|^{-\lambda}\operatorname{sgn} y$	(13.10)
$\exp(2\pi i a x)$	$\delta(y-a)$	(13.11)
$\exp(2\pi i a x)\operatorname{sgn} x$	$(\pi i)^{-1}(y-a)^{-1}$	(13.12)

Further pairs appear in (16.12)–(16.14).

hand member is absolutely integrable for $\varepsilon>0$,

$$\exp(-2\pi\varepsilon|x|)\operatorname{sgn} x \leftrightarrow \left(-\frac{i}{\pi}\right)\frac{y}{y^2+\varepsilon^2}.$$

On taking the limit as $\varepsilon\to0^+$ and using the representation in (12.27) we establish (13.6) for the case $m=1$, whence the cases $m\geqslant2$ follow on repeatedly differentiating the right hand functional.

Transform (13.7) is established from the following pair, in which the left member is absolutely integrable for $\varepsilon>0$:

$$[\ln(2\pi|x|)+C]\exp(-2\pi\varepsilon|x|)\leftrightarrow(-\tfrac{1}{2})\left[\frac{\varepsilon\ln(y^2+\varepsilon^2)+2y\arctan(y/\varepsilon)}{\pi(y^2+\varepsilon^2)}\right].$$

Here, as elsewhere, $-\pi/2<\arctan<\pi/2$, and C is Euler's constant,

$$C=\lim_{N\to\infty}\left[\left(\sum_{k=1}^{N}\frac{1}{k}\right)-\ln N\right]=0.5772\ldots. \qquad (13.13)$$

On taking the limit in S' as $\varepsilon\to0^+$ we establish (13.7) for $m=1$, where the value of $\psi(m)$ depends on the convention chosen for the value of B_1 in the definition of $y^{-1}\operatorname{sgn} y$, (12.35). Transform (13.7) for $m\geqslant2$ is then obtained by repeatedly differentiating the right hand member. If the functionals y^{-m} sgn y are defined according to (12.35) with the B_m given by (12.37) then it follows that

$$\psi(1) = -C, \qquad \psi(2) = 1 - C,$$
$$\psi(m) = -C + \sum_{k=1}^{m-1} \frac{1}{k} \qquad (m > 2). \tag{13.14}$$

If the B_m are left arbitrary, then $\psi(m)$ must be replaced by an arbitrary constant.

Transform (13.8) is established from the following pair, in which the left hand member is absolutely integrable for each $0 < \alpha < \infty$, and $\varepsilon > 0$:

$$x^{\alpha-1} \exp(-2\pi\varepsilon|x|)H(\pm x) \leftrightarrow \frac{\Gamma(\alpha)}{(2\pi)^\alpha} \frac{\exp[\mp i\alpha \arctan(y/\varepsilon)]}{(y^2+\varepsilon^2)^{\alpha/2}}.$$

On taking the limit in S' as $\varepsilon \to 0^+$ transform (13.8) follows using definition (12.34).

Transforms (13.9) and (13.10) may, for $0 < \lambda < 1$, be derived by starting from the pairs (9.2) and (9.3) already established as Fourier pairs in class K. On differentiating the right hand member m times and replacing $(\alpha + m)$ by λ we arrive at (13.9) and (13.10) using the definitions (12.31) and (12.32).

Finally other Fourier pairs may be generated by using the shift theorems, and for instance (13.11) and (13.12) follow from (13.5) and (13.6) on putting $n=0$, $m=1$, and replacing $F(y)$ by $F(y-a)$ in each case.

13.4 The convolution and product of functionals

A convolution product, $\tilde{f} * \tilde{g}$, between two functionals in S' is not defined for every pair of functionals, and for instance no meaning is attached to the convolution $\tilde{f} * \tilde{f}$ when $\tilde{f}(x) = f(x) = 1$ at all x. We now describe various conditions under which a convolution is defined, the simplest case arising when one of the functionals is regular and equal to a good function, there being in this case no restriction on the other functional.

If $\tilde{f} \in S'$ and $\varphi \in S$ we define the convolution of $\tilde{f}$ and φ as the *ordinary* function, say h, defined pointwise for all real a by

$$h(a) = \langle \tilde{f}, \varphi_a \rangle, \tag{13.15}$$

where φ_a is the good function $\varphi_a(x) = \varphi(a - x)$. This definition is consistent with previous definitions, and when $\tilde{f}$ is regular (13.15) becomes

$$h(a) = \int_{-\infty}^{+\infty} f(x)\varphi(a-x)\mathrm{d}x.$$

Accordingly we use the symbols $\tilde{f} * \varphi$ and $\varphi * \tilde{f}$ interchangeably to represent the convolution. $\tilde{f} * \varphi(x)$ is defined for all x and it can be shown that it is a *multiplier* in S' (see section 12.3) and is thus infinitely differentiable at all x.

We now have the following convolution theorem.

Theorem 13.5 Given $\tilde{f} \leftrightarrow \tilde{F}$ in S' and $\varphi \leftrightarrow \Phi$ in S, it follows that $\tilde{f}*\varphi \leftrightarrow \tilde{F}\Phi$ and $\tilde{f}\varphi \leftrightarrow \tilde{F}*\Phi$ are Fourier pairs in S'. □

The following results involving limits are also valid. If $\tilde{f}_n \to \tilde{f}$ in S' and $\varphi \in S$ then $\tilde{f}_n * \varphi \to \tilde{f}*\varphi$ in S'. Likewise if $\varphi_n \to \varphi$ in S and $\tilde{f} \in S'$, then $\tilde{f}*\varphi_n \to \tilde{f}*\varphi$ in S'. Note, however, that it does *not* necessarily follow that $\tilde{f}*\varphi_n$ will converge in S' to $\tilde{f}*\varphi$ when $\{\varphi_n\}$ is a sequence of good functions converging to φ in S' (rather than in S): the case $\tilde{f}=f(x)=1$, $\varphi(x)=0$ and

$$\varphi_n(x) = n^{-1}\exp(-\pi x^2/n^2) \tag{13.16}$$

provides a counterexample to demonstrate this, since as $n \to \infty$ we find $\tilde{f}*\varphi_n \to 1$ in S' whilst $\tilde{f}*\varphi = 0$ and $\varphi_n \to \varphi$ in S'.

Given $\tilde{f} \in S'$, $\varphi \in S$ and $\psi \in S$ it will follow that $(\tilde{f}*\varphi)\psi$ and $(\tilde{f}\varphi)*\psi$ are each in S, and this opens the way to constructing a sequence of good functions converging in S' to $\tilde{f}$.

Theorem 13.6 Suppose that $\tilde{f} \in S'$, $\varphi \in S$, $\int \varphi = 1$, $\psi \in S$ and $\psi(0)=1$: then it follows that

$$(\tilde{f}*\varphi_n)\psi_n \to \tilde{f} \quad \text{in } S'$$
and
$$(\tilde{f}\psi_n)*\varphi_n \to \tilde{f} \quad \text{in } S', \tag{13.17}$$

where for $n=1,2,3,\ldots$ $\varphi_n(x) = n\varphi(nx)$ and $\psi_n(x) = \psi(x/n)$. □

We now extend the definition of the convolution operation by introducing a class of functionals called convolutes. A functional $\tilde{f} \in S'$ is said to be a *convolute* in S' if, for each $\varphi \in S$, $\tilde{f}*\varphi$ is in class S. As examples the delta functional and each of its derivatives is a convolute in S', as also is any functional in S' of bounded support (such as the rectangle function), and any regular functional equal to a good function. The functionals x^{-1}, $(x+\mathrm{i}0)^{-1}$, x^2, and $\ln|x|$ are, however, not convolutes. It can be shown that if $\tilde{f}$ is a convolute, then so also is its transpose $\tilde{f}_{\mathrm{T}}$, and also $x^n\tilde{f}$ and $\tilde{f}^{(n)}$ will be convolutes, for $n = 1, 2, 3, \ldots$.

The following theorem now allows us to define a convolution product between an arbitrary functional in S' and a convolute in S'.

Theorem 13.7 Suppose $\tilde{f} \in S'$ and that $\tilde{g}$ is a convolute in S': then there exists a functional in S', which we write as $\tilde{f}*\tilde{g}$ or as $\tilde{g}*\tilde{f}$, and call the convolute of $\tilde{f}$ and $\tilde{g}$, such that for each $\varphi \in S$

$$\langle \tilde{f}*\tilde{g}, \varphi \rangle = \langle \tilde{f}, \tilde{g}_{\mathrm{T}}*\varphi \rangle, \tag{13.18}$$

where $\tilde{g}_{\mathrm{T}}$ is the transpose of $\tilde{g}$. □

This definition is consistent with the previous definition of a convolution in S', and (13.18) is a plausible extension of the following equation which is valid for good functions φ, ψ and θ:

$$\int (\varphi * \psi)\theta = \int \varphi(\psi_T * \theta),$$

where ψ_T is the transpose of ψ. As examples of this kind of convolution product, we have $\tilde{\delta} * \tilde{\delta} = \tilde{\delta}$, and for each $\tilde{f} \in S'$, $\tilde{\delta} * \tilde{f} = \tilde{f}$, and for $n = 1, 2, 3, \ldots,$

$$\tilde{\delta}^{(n)} * \tilde{f} = \tilde{f}^{(n)}.$$

An alternative, and equivalent, definition of $\tilde{f} * \tilde{g}$, when $\tilde{g}$ is convolute, may be based on the fact that $f_n * \tilde{g} \to \tilde{f} * \tilde{g}$ in S', where $\{f_n\}_{n=1}^{\infty}$ is any sequence of functions in S converging to $\tilde{f}$ in S'. The existence of such a sequence is guaranteed by theorem 13.6. Note that it does *not* necessarily follow that $\tilde{f} * \tilde{g}_n$ will converge in S' to $\tilde{f} * \tilde{g}$ when $\{g_n\}_{n=1}^{\infty}$ is a sequence of good functions converging to the convolute $\tilde{g}$ in S': the counterexample based on (13.16) provided an example of this. However, it is true that if $\tilde{g}$ is a convolute in S' then $\tilde{f}_n * \tilde{g} \to \tilde{f} * \tilde{g}$ in S' whenever $\tilde{f}_n \to \tilde{f}$ in S'.

The multipliers in S' and the convolutes in S' are related by Fourier transformation, in that a functional in S' will be convolute in S' if, and only if, its Fourier transform is a multiplier in S'. Moreover for each $\tilde{f} \in S'$ and each $\varphi \in S$, it can be shown that $\tilde{f} * \varphi$ is a multiplier in S' and that $\tilde{f}\varphi$ is a convolute in S'. This leads to the following convolution theorem.

Theorem 13.8 Suppose $\tilde{f} \leftrightarrow \tilde{F}$ in S', and $\tilde{g} \leftrightarrow \tilde{G}$ in S': then it follows that (i) whenever $\tilde{g}$ is a convolute

$$\tilde{f} * \tilde{g} \leftrightarrow \tilde{F}\tilde{G},$$

and (ii) whenever $\tilde{g}$ is a multiplier,

$$\tilde{f}\tilde{g} \leftrightarrow \tilde{F} * \tilde{G}. \quad \square$$

As an example we may use this theorem to evaluate $x^{-1} * r(x)$, where $r(x) = 1$ $(|x| < 1)$, $= 0$ $(|x| > 1)$. From (8.5) and (13.6) and theorem 13.8 we have

$$x^{-1} * r(x) \leftrightarrow (-i\pi \operatorname{sgn} y)\left(\frac{\sin 2\pi y}{\pi y}\right).$$

The right hand side of this is regular and in L^2, and its inverse transform is obtained from (8.8). Whence we have the generalized equation

$$x^{-1} * \operatorname{rect}(x) = \operatorname{sgn} x \ln\left|\frac{|x|+1}{|x|-1}\right|.$$

This equation is also valid as an ordinary convolution when $|x| > 1$. As another example the convolution $\tilde{\delta}*\tilde{\delta}=\tilde{\delta}$ is readily established using theorem 13.8 and the pair $\tilde{\delta}\leftrightarrow 1$.

We now consider the convolution $\tilde{f}*\tilde{g}$ when neither $\tilde{f}$ nor $\tilde{g}$ is a convolute, and consider the product $\tilde{f}\tilde{g}$ when neither $\tilde{f}$ nor $\tilde{g}$ is a multiplier. It is necessary to place constraints upon both $\tilde{f}$ and $\tilde{g}$ in each case. Different authors adopt different approaches, and we present the following approach as one that is natural in many practical applications. A product is defined as follows.

Definitions Let $\tilde{f}$ and $\tilde{g}$ be functionals in S' and suppose that for each $\varphi\in S$ satisfying $\int\varphi=1$ the sequences $\tilde{g}(\tilde{f}*\varphi_n)$ and $\tilde{f}(\tilde{g}*\varphi_n)$ converge in S' to a common limit functional in S', where $\varphi_n(x)=n\varphi(nx)$, the limit functional being independent of the choice of φ: then the limit function is defined as the product in S' of $\tilde{f}$ and $\tilde{g}$ and is written equivalently as $\tilde{f}\tilde{g}$ or as $\tilde{g}\tilde{f}$. □

A convolution is defined analogously, as follows.

Definition Let $\tilde{f}$ and $\tilde{g}$ be functionals in S' and suppose that for each $\psi\in S$ satisfying $\psi(0)=1$, the sequences $\tilde{f}*(\tilde{g}\psi_n)$ and $(\tilde{f}\psi_n)*\tilde{g}$ converge in S' to a common limit functional, where $\psi_n(x)=\psi(x/n)$, the limit functional being independent of the choice of ψ: then the limit functional is defined as the convolution of $\tilde{f}$ and $\tilde{g}$ and is written equivalently as $\tilde{f}*\tilde{g}$ or as $\tilde{g}*\tilde{f}$. □

These definitions are consistent with previous definitions, when, say, $\tilde{g}$ is a multiplier or a convolute as appropriate. The definitions are however more general in that meanings are now, for instance, attached to the following products, $[\tilde{\delta}(x-1)][\operatorname{sgn} x]$, $x^{-1}\exp(-|x|)$, $[\tilde{\delta}(x)][\tilde{\delta}(x-1)]$, and meanings are attached to the convolutions $x^{-2}*x^{-3}$, $x^{-1}*(1+x^2)^{-1}$. Not all pairs can be multiplied or convoluted, however, and for instance the above definitions attach no meaning to $x*x$, or to the products $x^{-1}x^{-1}$, $\tilde{\delta}\tilde{\delta}$ or $\operatorname{sgn} x\delta(x)$.

The convolution theorem may now be generalized still further as follows.

Theorem 13.9 Suppose $\tilde{f}\leftrightarrow\tilde{F}$ in S' and $\tilde{g}\leftrightarrow\tilde{G}$ in S': then (i) $\tilde{f}\tilde{g}$ will exist if, and only if, $\tilde{F}*\tilde{G}$ exists, and (ii) $\tilde{f}*\tilde{g}$ will exist, if and only if, $\tilde{F}\tilde{G}$ exists, and (iii) it will follow that

$$\tilde{f}*\tilde{g}\leftrightarrow\tilde{F}\tilde{G} \text{ in } S'$$

whenever these quantities exist, and (iv)

$$\tilde{f}\tilde{g}\leftrightarrow\tilde{F}*\tilde{G} \text{ in } S'$$

whenever these quantities exist. □

Differentiation of these products and convolutions leads to the following theorems.

Theorem 13.10 Let $\tilde{f}$ and $\tilde{g}$ be functionals in S' and suppose that $\tilde{f}\tilde{g}$ and $\tilde{f}\tilde{g}'$ are defined: then $\tilde{f}'\tilde{g}$ will also be defined and it will follow that

$$(\tilde{f}\tilde{g})' = \tilde{f}'\tilde{g} + \tilde{f}\tilde{g}'. \quad \square \tag{13.19}$$

Theorem 13.11 Let $\tilde{f}$ and $\tilde{g}$ be functionals in S' and suppose that $\tilde{f}*\tilde{g}$ is defined: then it will follow that $\tilde{f}*\tilde{g}'$ and $\tilde{f}'*\tilde{g}$ are defined and

$$(\tilde{f}*\tilde{g})' = (\tilde{f}'*\tilde{g}) = (\tilde{f}*\tilde{g}'). \quad \square \tag{13.20}$$

Although generalized products and convolutions are, by definition, commutative, they are not always associative and $(\tilde{f}\tilde{g})\tilde{h}$ is not necessarily the same as $\tilde{f}(\tilde{g}\tilde{h})$. For instance, we may consider the triple products between x^{-1}, x and δ. We have

$$(x^{-1}x)\delta = 1 \times \delta = \delta \tag{13.21}$$

and

$$x^{-1}(x\delta) = x^{-1} \times 0 = 0 \tag{13.22}$$

whilst $(x^{-1}\tilde{\delta})x$ is undefined since $x^{-1}\tilde{\delta}$ is undefined. However, if $\tilde{f}$, $\tilde{g}$ and $\tilde{f}\tilde{g}$ exist in S' and h is a multiplier in S' then the product *is* associative and $\tilde{f}(\tilde{g}h) = (\tilde{f}\tilde{g})h = (h\tilde{f})\tilde{g}$. Similarly, although $(\tilde{f}*\tilde{g})*\tilde{h}$ may differ from $\tilde{f}*(\tilde{g}*\tilde{h})$, the triple convolution becomes independent of the order and bracketing when one of the three is a convolute, and the convolute of the remaining pair is defined. For example, on Fourier transforming (13.21) and (13.22) we obtain

$$(\text{sgn}*\delta')*1 = 2\delta*1 = 2 \tag{13.23}$$

and

$$\text{sgn}*(\delta'*1) = \text{sgn}*0 = 0 \tag{13.24}$$

whilst $(\text{sgn}*1)*\delta'$ is undefined since $\text{sgn}*1$ is undefined.

Products and convolutions in S' are distributive in the sense that if $\tilde{f}\tilde{g}$ and $\tilde{f}\tilde{h}$ are defined, then so also will $\tilde{f}(\tilde{g}+\tilde{h})$ and

$$\tilde{f}(\tilde{g}+\tilde{h}) = \tilde{f}\tilde{g} + \tilde{f}\tilde{h}.$$

Similarly, when $\tilde{f}*\tilde{g}$ and $\tilde{f}*\tilde{h}$ are defined, so also will $\tilde{f}*(\tilde{g}+\tilde{h})$ and it will follow that

$$\tilde{f}*(\tilde{g}+\tilde{h}) = \tilde{f}*\tilde{g} + \tilde{f}*\tilde{h}.$$

Certain conditions are sufficient to ensure the existence of $\tilde{f}*\tilde{g}$. For instance, if $\tilde{f}=f$ and $\tilde{g}=g$ are regular functionals in S', and if the ordinary convolution $|f|*|g|$ exists and defines a regular functional in S', then $\tilde{f}*\tilde{g}$ will exist and will equal the ordinary convolution $f*g$; this will be so if $\tilde{f}$ and $\tilde{g}$ are in class K and satisfy condition (CX) of theorem 9.6 (or one of conditions (CX1), (CX2) and (CX3) in theorem 9.7). Also, the convolution

$\tilde{f} * \tilde{g}$ will exist if $\tilde{f}$ and $\tilde{g}$ are causal functionals in S'; more generally, if $\tilde{f} \in S'$ and $\tilde{g} \in S'$ are such that $\tilde{f}=0$ on $(-\infty, a)$ and $\tilde{g}=0$ on $(-\infty, b)$, then $\tilde{f} * \tilde{g}$ exists in S' and will equal zero on $(-\infty, c)$, where $c=a+b$ for any finite real numbers a and b.

The following conditions are sufficient to ensure the existence of the product $\tilde{F}\tilde{G}$. If F and G are ordinary functionals in class K that satisfy condition (CY) of theorem 9.6 (or one of conditions (CY1), (CY2) and (CY3) of theorem 9.8) then F, G and the ordinary product FG each define regular functionals in S', say $\tilde{F}$, $\tilde{G}$ and $\tilde{H}$, and it will follow that $\tilde{F}\tilde{G}$ is defined and equals $\tilde{H}$. Another condition depends upon $\tilde{F}$ and $\tilde{G}$ not 'overlapping': if $\tilde{F}$ and $\tilde{G}$ are in S' and if $\tilde{F}=0$ on $(-\infty, b)$ and $\tilde{G}=0$ on (a, ∞) for some finite values $a<b$, then the product $\tilde{F}\tilde{G}$ exists and equals zero. However, if $\tilde{F}$ and $\tilde{G}$ are regular in S' and the ordinary product FG defines a regular functional in S', it does *not* follow that $\tilde{F}\tilde{G}$ will necessarily be defined. A counterexample is provided by

$$F(y)=\begin{cases}|y|^{-2/3}, & y>0\\ 0, & y\leqslant 0\end{cases} \tag{13.25}$$

$$G(y)=\begin{cases}0, & y\geqslant 0\\ |y|^{-2/3}, & y<0.\end{cases} \tag{13.26}$$

In this case the ordinary product is zero at all y, and F and G are in class K and so define regular functionals in S', yet $\tilde{F}\tilde{G}$ is undefined since the sequence $\tilde{F}(\tilde{G} * \varphi_n)$, which appears in the general definition of the product, fails to converge for all good φ satisfying, $\int \varphi=1$ and $\varphi_n(y)=n\varphi(ny)$.

We conclude with a special case in which the product definition gives a result which is a natural extension of (12.21). If a functional $\tilde{f}$ in S' is equal to an ordinary function f on some interval $(a-\varepsilon, a+\varepsilon)$, where a is real and $\varepsilon>0$, and if $f(x)$ is continuous at $x=a$, then the product $\tilde{\delta}(x-a)\tilde{f}(x)$ will exist, and it will follow that

$$\tilde{\delta}(x-a)f(x)=f(a)\tilde{\delta}(x-a);$$

moreover if f is n times differentiable at $x=a$, then it will follow that

$$\tilde{f}(x)\tilde{\delta}^{(n)}(x-a)=\sum_{p=0}^{n}\frac{(-1)^p n!\, f^{(p)}(a)}{p!\,(n-p)!}\tilde{\delta}^{(n-p)}(x-a). \tag{13.27}$$

14

Fourier transformation of generalized functions – II

14.1 Functionals of types D' and Z'

In order to give meaning to the Fourier transform of any locally integrable function it is necessary to generalize beyond the functionals in S', and we describe now the functionals in classes D' and Z' which provide the necessary generalization. We give examples in section 14.2, but start in this section with the basic concepts.

In summary, the functionals of class D' are based on the use of test functions in class D (the good functions of bounded support), whilst the functionals in class Z' are based on the use of test functions in class Z (the Fourier transforms of functions in D). Much, but not all, of chapter 12 can simply be adapted by replacing S by D (or Z) and S' by D' (or Z'). For instance, functionals of type D' and Z' are defined as follows.

Definition An association of exactly one real or complex number with each $\varphi \in D$ is said to be a *functional in class* D' if there exists at least one sequence $\{f_n\}_{n=1}^{\infty}$ of ordinary functions such that $f_n\varphi \in L$ for each $\varphi \in D$ and each n, and such that for each $\varphi \in D$ the number associated with φ is equal to $\lim_{n\to\infty} \int f_n\varphi$. □

Likewise:

Definition An association of exactly one number with each $\varphi \in Z$ is said to be a *functional in class* Z' if there exists at least one sequence $\{f_n\}_{n=1}^{\infty}$ of ordinary functions such that $f_n\varphi \in L$ for each $\varphi \in Z$ and each n, and such that for each $\varphi \in Z$ the number associated with φ is equal to $\lim_{n\to\infty} \int f_n\varphi$. □

The names *distribution*, *tempered distribution* and *ultradistribution* are often used to describe functionals in D', S' and Z', respectively, whilst the term *generalized function* is used to cover all three classes as well as other variants introduced by Gel'fand *et al.* (1964–9).

The definitions of convergence in D' and convergence in Z' of sequences of functions are obtained from the definition of convergence in S', (12.4), simply by reading D (or Z) for S and D' (or Z') for S'. However, on proceeding through section 12.2 in this way there are places where more radical changes are necessary, and we now describe these in turn.

If f is an ordinary function, then the requirement that $f\varphi \in L$ for each $\varphi \in S$ is equivalent to the conditions that $f \in K$ (section 9.2), so that the regular functionals in S' correspond to functions in K. However, the requirement that $f\varphi \in L$ for each $\varphi \in D$ is equivalent to the condition that $f \in L_{\text{LOC}}$, so that the regular functionals in D' correspond to the class of functions in L_{LOC}. For example, the ordinary function $f(x) = \exp(x)$ corresponds to a regular functional in D', but to no functional in S'.

Alternative and equivalent definitions of functionals in D' and Z' are possible, based on the idea of a linear continuous functional. A functional defined on D or on Z is defined to be linear if (12.6) is valid for all complex α and β, just as for the functionals defined on S. However, continuity requires a new definition and is now based on the following definitions of convergence in D and convergence in Z.

Definition A sequence $\{\varphi_n\}_{n=1}^{\infty}$ of functions $\varphi_n \in D$ is said to converge in D if each of the φ_n is equal to zero outside of some fixed interval $[a, b]$, and if φ_n and each of its derivatives converges uniformly on $[a, b]$ as $n \to \infty$. □

Convergence in Z is more involved, and its definition relies on the fact that each $\varphi(x)$ in Z, defined on the real axis, may be analytically extended as described in section 7.3 to a function $\varphi_c(z)$ defined on the complex plane.

Definition A sequence $\{\varphi_n\}_{n=1}^{\infty}$ of functions $\varphi_n \in Z$ is said to converge in Z if $\varphi_n(x)$ converges uniformly on every finite interval of the x-axis as $n \to \infty$, and if in addition for each $k = 1, 2, 3, \ldots$ and each complex value of z,

$$|z^k \varphi_{cn}(z)| < C_k \exp(b|\text{Im } z|), \tag{14.1}$$

where $\varphi_{cn}(z)$ is the analytic extension of $\varphi_n(x)$, b is a positive constant independent of n, k or z, and C_k is a positive constant dependent on k but independent of n or z. □

Convergent sequences in D (or Z) necessarily converge uniformly on $(-\infty, \infty)$ to a limit function that is in D (or Z).

Continuity of functionals defined on D (or Z) may now be defined by analogy with the definition of continuity of functionals on S, using (12.7) and replacing S by D (or Z). An alternative method of defining functionals in D' (or Z') is now possible through the following.

Theorem 14.1 The class of functionals in D' is identical to the class of linear continuous functionals on D, and likewise the class of functionals in Z' is identical to the class of linear continuous functionals on Z. □

Every functional in D' (or Z') can be represented as the limit in D' (or Z') of a sequence of functions in D (or Z), so that the definitions of functionals in D' and Z' given at the beginning of this section can equivalently be based on the existence of at least one such sequence.

Regular functionals in D' (or Z') are defined using (12.8), on replacing S by D (or Z) and S' by D' or (Z'). Likewise functionals in D' (or Z') that are not regular are called singular. Similarly a singular functional in D' may be regular on part of the x-axis, by analogy with the discussion based on (12.9). However, a singular functional in Z', that is not also in either D' or S', *cannot* be defined as regular on some finite or semi-infinite interval of the x-axis since the test functions in Z have zero value on at most a null set. For the same reason one does not define equality of functionals in Z' over finite intervals of the x-axis. The concept of functionals of bounded support, or of causality, is not defined for functionals in Z', unless they are also in D' or S'.

Many of the basic definitions and theorems governing functionals in D' and in Z' are closely analogous to those governing functionals in S'. Indeed section 12.3 is written in such a way that it becomes applicable to functionals in D' if throughout class S is replaced by class D, and class S' is replaced by class D'; likewise section 12.3 (apart from the paragraph based on (12.9)) becomes applicable to functionals in Z' if S is replaced by Z and S' by Z'. In this way the concepts of equality, addition and subtraction, differentiation, multiplication by a constant, scaling, transposition, complex conjugation and the integration of functionals in D' and in Z' are described. Likewise also the definition and implications of convergence in D' (or Z') of a sequence of functionals in D' (or Z') are described. Also multipliers in D' (or Z'), and the product of a functional in D' (or Z') with a multiplier in the same class are defined. The examples in section 12.3 also remain valid when S is replaced by D (or Z) and S' is replaced by D' (or Z'), since a functional in S' is necessarily also a functional in D' and in Z'.

We conclude with a few results which are *not* common to all three classes of functional, starting with the explicit characterization of a multiplier.

Theorem 14.2 An ordinary function $f(x)$ will be a multiplier in S' if, and only if, $f(x)$ is infinitely differentiable at all real x and if f and each derivative is bounded by a polynomial, the polynomial being not necessarily the same one for each derivative. □

Theorem 14.3 An ordinary function $f(x)$ will be a multiplier in D' if, and only if, $f(x)$ is infinitely differentiable at each real value of x. □

The characterization of a multiplier in Z' is less easy and relies on the idea of analytic extension over the complex plane.

Theorem 14.4 An ordinary function $f(x)$ will be a multiplier in Z' if, and only if, there exists a function $f_c(z)$ defined and differentiable everywhere on the complex plane (i.e. an entire function) such that for all real x $f(x)=f_c(x)$ and such that at each complex z,

$$|f_c(z)| \leqslant C\exp(b|\text{Im } z|)(1+|z|^q) \tag{14.2}$$

where C, b and q are positive constants independent of z. □

Another way of generating multipliers in Z' follows from:

Theorem 14.5 A function $f(x)$ will be a multiplier in Z' if, and only if, it is the Fourier transform of some functional in S' of bounded support. □

Thus the functions $(\sin x)/x$ and $\exp(-2\pi ix)$ are multipliers in Z', being the Fourier transforms of a rectangle function and of $\delta(x-1)$, respectively. The function $\exp(x)$ is a multiplier in D' but not in S' or Z'.

Another contrast between the three types of functional concerns the effect of repeated integration. Repeated integration of a functional in S' inevitably leads eventually to a regular functional that is everywhere continuous. For example, the delta function is the second derivative of the continuous function $f(x)=|x|/2$. The same is not necessarily true, however, for functionals in D' or in Z', though an analogous result holds *locally* for functionals in D'. The following theorems express this more precisely.

Theorem 14.6 Let $\tilde{f}$ be a functional in S': then there exists a positive integer n and an ordinary function $f_0(x)$, that is continuous at all x and whose modulus is bounded by some polynomial, such that $\tilde{f}$ is the nth generalized derivative of f_0. □

Theorem 14.7 Let $\tilde{f}$ be a functional in D' and let $[a, b]$ be some fixed finite interval on the real axis: then there exists a positive integer n and an ordinary function $f_0(x)$, that is continuous at all x, such that

$$\langle \tilde{f}, \varphi \rangle = \langle \tilde{f}_0^{(n)}, \varphi \rangle$$

whenever $\varphi \in D$ has the value zero outside of the interval $[a, b]$. □

As an example we may choose the functional $\tilde{f} \in D'$ defined as follows, for each $\varphi \in D$,

$$\langle \tilde{f}, \varphi \rangle = \varphi(0) + \varphi'(1) + \varphi''(2) + \varphi'''(3) + \cdots, \tag{14.3}$$

or equivalently

$$\tilde{f} = \lim_{N\to\infty} \sum_{n=0}^{N} (-1)^n \tilde{\delta}^{(n)}(x-n) \quad (\text{in } D').$$

This functional is not in class S', and it is not possible to express $\tilde{f}$ in the form $f_0^{(m)}$ for any continuous function f_0 or any m unless we restrict attention to some finite interval of the x-axis. As another example, consider the limit

$$\lim_{N\to\infty}\left[\sum_{n=1}^{N} x^n \cos(2\pi n x)\right]. \tag{14.4}$$

This *does* converge in Z' to a functional in Z', say $\tilde{f}$. This functional $\tilde{f}$ is, however, not in D' or in S' and it cannot be expressed as the repeated derivative of some continuous function (even over a finite interval).

14.2 Fourier transformation of functionals in D'

The Fourier transformation of a functional in D' relies for its definition on the following theorem, the transform being a functional in Z'.

Theorem 14.8 Let $\tilde{f}$ be a functional in D': then there exists a functional in Z', say $\tilde{F}$, such that for each $\Phi\in Z$

$$\left.\begin{aligned}&\langle\tilde{F},\Phi\rangle=\langle\tilde{f},\varphi_{\mathrm{T}}\rangle\\ \text{and}\quad&\langle\tilde{F},\Phi^*\rangle=\langle\tilde{f},\varphi^*\rangle\end{aligned}\right\} \tag{14.5}$$

where $\varphi\in D$ is the inverse transform of Φ, and φ_{T} is the transpose of φ. □

When the conditions of this theorem are satisfied we define $\tilde{F}$ as the Fourier transform of $\tilde{f}$, $\tilde{f}$ as the inverse transform of $\tilde{F}$, and we refer to $\tilde{f}$ and $\tilde{F}$ as a Fourier pair in D' writing $\tilde{f}\leftrightarrow\tilde{F}$. The definition is unique in that $\tilde{f}\leftrightarrow\tilde{F}$ and $\tilde{f}\leftrightarrow\tilde{G}$ together imply that $\tilde{F}=\tilde{G}$, whilst $\tilde{f}\leftrightarrow\tilde{F}$ and $\tilde{g}\leftrightarrow\tilde{F}$ together imply that $\tilde{f}=\tilde{g}$. The Fourier transform of a functional $\tilde{f}$ in class Z' is likewise defined as the functional $\tilde{F}\in D'$ satisfying (14.5) for every pair $\varphi\leftrightarrow\Phi$ in which φ is in class Z: we then write $\tilde{f}\leftrightarrow\tilde{F}$ and call it a Fourier pair in Z'. These definitions are consistent with the previous, and less general, definition of Fourier pairs in S'. Each of the following theorems on Fourier pairs in D' has its counterpart for Fourier pairs in Z'.

Convergence of a sequence of functionals in D' implies convergence in Z' of the transform, and vice versa, as follows.

Theorem 14.9 A sequence $\{f_n\}_{n=1}^{\infty}$ of functionals in D' will converge if, and only if, the sequence $\{F_n\}_{n=1}^{\infty}$ of Fourier transforms converge in Z'; moreover when these sequences converge in this way then the limit functionals $\tilde{f}\in D'$ and $\tilde{F}\in Z'$ form a Fourier pair. □

The above theorem is clearly the analogue of theorem 13.2 which applies to functionals in S'. Likewise the results on linearity, shifting, scaling, transposition and complex conjugation which were described in section 13.2 for Fourier pairs in S' have their exact analogues for functionals $\tilde{f}$ and

$\tilde{g}$ in D' with transforms $\tilde{F}$ and $\tilde{G}$ in Z'. The differentiation theorem now runs as follows.

Theorem 14.10 Suppose $\tilde{f}\in D'$ has the Fourier transform $\tilde{F}\in Z'$ and that $\tilde{f}^{(n)}$ and $\tilde{F}^{(n)}$ are the nth differentials of $\tilde{f}$ and $\tilde{F}$: then it will follow that for each $n=0,1,2,3,\ldots$

$$\tilde{f}^{(n)}(x)\leftrightarrow(2\pi \mathrm{i}y)^n\tilde{F}(y)$$

$$(-2\pi \mathrm{i}x)^n\tilde{f}(x)\leftrightarrow\tilde{F}^{(n)}(y),$$

where $\tilde{f}^{(n)}$ is in D' and $\tilde{F}^{(n)}$ is in Z'. □

We now give examples of functionals in D' with their Fourier transforms. Since functionals in S' are always also functionals in D' and in Z' it follows that all the Fourier pairs given in section 13.3 are also examples of transforms of functionals in D'; we now concentrate therefore on examples of functionals in D' that are not also functionals in S'.

The functionals $\tilde{f}\in D'$ and $\tilde{F}\in Z'$ form a Fourier pair, where $\tilde{f}$ and $\tilde{F}$ are defined by:

$$\tilde{f}(x)=\lim_{N\to\infty}\sum_{n=1}^{N}\tfrac{1}{2}[\delta^{(n)}(x-n)+\delta^{(n)}(x+n)] \tag{14.6}$$

$$\tilde{F}(y)=\lim_{N\to\infty}\sum_{n=1}^{N}[(2\pi \mathrm{i}y)^n\cos(2\pi yn)]. \tag{14.7}$$

We may note that when the two summations are terminated at a finite value of N, there results a Fourier pair in S'. However, the series defining $\tilde{f}$ converges in D' but not in S', whilst that defining $\tilde{F}$ converges in Z' but not in S'.

As a second example the following functionals $\tilde{f}\in D'$ and $\tilde{F}\in Z'$ form a Fourier pair, where $\tilde{f}$ is regular and equal to the ordinary function $\exp(-2\pi ax)$, for real a, and $\tilde{F}$ is defined by the requirement that for each $\Phi\in Z$

$$\langle\tilde{F},\Phi\rangle=\sum_{n=0}^{\infty}\frac{(\mathrm{i}a)^n\Phi^{(n)}(0)}{n!}. \tag{14.8}$$

There are other ways of writing this functional $\tilde{F}$. Since Φ is an entire function we may utilize a Taylor expansion, (7.23), to analytically extend $\Phi(y)$ into $\Phi_c(z)$. On comparison with (14.8) above we see that

$$\langle\tilde{F},\Phi\rangle=\Phi_c(\mathrm{i}a). \tag{14.9}$$

Since $\tilde{F}$ 'picks out' the value of $\Phi_c(z)$ at $z=\mathrm{i}a$, the suggestive symbol $\delta(z-\mathrm{i}a)$ is often used to represent $\tilde{F}$.

As a third example we consider the ordinary, continuous functions defined by the following pointwise convergent series:

$$f(x) = \sum_{n=1}^{\infty} 2^n \cos(2\pi 2^n x)\{\exp[-\pi(x-n)^2] + \exp[-\pi(x+n)^2]\}$$

$$F(y) = \sum_{n=1}^{\infty} 2^n \cos(2\pi n y)\{\exp[-\pi(y-2^n)^2] + \exp[-\pi(y+2^n)^2]\}.$$

The functions f and F define regular functionals $\tilde{f}$ and $\tilde{F}$ in D' and Z', respectively, and $\tilde{f} \leftrightarrow \tilde{F}$ in D'. In fact the function F is in class K, so that $\tilde{F} \in S'$, and consequently $\tilde{f}$ is also in S'. However, interestingly, f is not in class K so that (theorem 9.1) $\tilde{f}$ is not a regular functional in S', despite its close relationship with f.

As a fourth example we consider the Fourier transform of the regular functional $\tilde{f}$ in D' equal to the ordinary function $\exp(\pi x^2)$. The Fourier transform of $\tilde{f}$ is the functional $\tilde{F}$ in Z' defined by the requirement that for each $\Phi \in Z$, with analytic extension $\Phi_c(z)$,

$$\langle \tilde{F}, \Phi \rangle = -\mathrm{i} \int_{-\mathrm{i}\infty}^{+\mathrm{i}\infty} \exp(\pi z^2)\Phi_c(z)\mathrm{d}z$$

$$= \int_{-\infty}^{\infty} \exp(-\pi\alpha^2)\Phi_c(\mathrm{i}\alpha)\mathrm{d}\alpha. \qquad (14.10)$$

In these two equivalent integrals z is a complex variable, the associated integration being along the imaginary axis, whilst α is a real variable.

In the above fourth example, $\tilde{F}$ is a so-called analytic functional in Z', Gel'fand *et al.* (1964–9), this class of functional being defined as follows. A functional $\tilde{F} \in Z'$ is called an *analytic functional in* Z' if there exists an ordinary function $F_0(z)$ and a finite or infinite contour C in the complex plane, such that for each $\Phi \in Z$,

$$\langle \tilde{F}, \Phi \rangle = \int_C F_0(z)\Phi_c(z)\mathrm{d}z \qquad (14.11)$$

where $\Phi_c(z)$ is the analytic extension of Φ. To specify an analytic functional in Z' it is thus sufficient to specify the ordinary function $F_0(z)$ and the contour C. The following theorem represents a kind of Fourier inversion theorem when $\tilde{F}$ is analytic in Z'.

Theorem 14.11 Suppose $\tilde{F}$ is an analytic functional in Z', specified by $F_0(z)$ and the finite or infinite contour C as in (14.11), and suppose also that $F_0(z)\exp(b \operatorname{Im} z)$ is absolutely integrable along C for every real b: then it follows that $\tilde{F}$ is the Fourier transform of the regular functional $\tilde{f} = f(x)$ in class D' where, at all x,

$$f(x) = \int_C F_0(z)\exp(2\pi \mathrm{i} x z)\mathrm{d}z. \quad \square \qquad (14.12)$$

On choosing $F_0(z) = \exp(\pi z^2)$ together with the contour C running along the imaginary axis from $z=-i\infty$ to $+i\infty$, it will be found that (14.12) can be evaluated using standard integrals to yield $f(x)=\exp(\pi x^2)$ in agreement with the example based on (14.10) above.

14.3 Transformation of products and convolutions in D'

The Fourier transformation of products and convolutions in D' and Z' is closely analogous to that for functionals in S', already described in section 13.4, and we now summarize the results without further explanation. Throughout it is to be understood that functionals of different types are not to be mixed, so that in a product $\tilde{f}\tilde{g}$ or a convolution $\tilde{f}*\tilde{g}$ both $\tilde{f}$ and $\tilde{g}$ must be in D' or both must belong to Z'.

Given $\tilde{f}\in D'$ (or Z') and $\varphi\in D$ (or Z) we define the convolution $\tilde{f}*\varphi$ as an ordinary function such that pointwise at each real a

$$\tilde{f}*\varphi(a)=\langle\tilde{f},\varphi_a\rangle$$

where φ_a is defined by $\varphi_a(x)=\varphi(a-x)$ at each a. We then find that

$$\tilde{f}_n*\varphi\rightarrow\tilde{f}*\varphi \text{ in } D' \quad (\text{or } Z')$$

whenever $\tilde{f}_n\rightarrow\tilde{f}$ in D' (or Z') and $\varphi\in D$ (or Z). Likewise

$$\tilde{f}*\varphi_a\rightarrow\tilde{f}*\varphi \text{ in } D' \quad (\text{or } Z')$$

whenever $\varphi_n\rightarrow\varphi$ in D (or in Z) and $\tilde{f}\in D'$ (or Z').

We say $\tilde{f}\in D'$ (or Z') is a *convolute* in D' (or Z') if $\tilde{f}*\varphi\in D$ (or Z) for each $\varphi\in D$ (or Z). A test function $\varphi\in D$ (or Z) is both a multiplier and a convolute in D (or Z). A functional $\tilde{f}\in D'$ will be a convolute in D' if, and only if, $\tilde{f}$ is equal to zero both on $(-\infty,-a)$ and on (a,∞) for some $a>0$. A correspondingly simple characterization of a convolute in Z' is not possible: however, the condition that $\tilde{F}$ is a convolute in Z' is equivalent to the condition that its inverse transform $\tilde{f}$ shall be a multiplier in D', see theorem 14.3. Likewise the condition that $\tilde{f}$ is a convolute in D' is equivalent to the condition that its transform $\tilde{F}$ is a multiplier in Z', theorem 14.4. If $\tilde{f}$ is a multiplier in D' (or Z') and $\varphi\in D$ (or Z), then $\tilde{f}\varphi$ is a convolute in D' (or Z'): likewise if $\tilde{f}$ is a convolute in D' (or Z') and $\varphi\in D$ (or Z), then $\tilde{f}*\varphi$ is a multiplier in D' (or Z'). If $\tilde{f}$ is a multiplier or convolute in D' or Z' (i.e. four categories) then $x^n\tilde{f}^{(m)}$ will be in the same category as $\tilde{f}$ for each $n=0,1,2,\ldots$ and each $m=0,1,2,3,\ldots$.

If $\tilde{f}\in D'$ (or Z') and φ and ψ are each in D (or each in Z) then it will follow that $(\tilde{f}*\varphi)\psi\in D$ (or Z) and that $(\tilde{f}\psi)*\varphi\in D$ (or Z): if in addition $\int\varphi=1$, $\psi(0)=1$, and if we introduce, for $n=1,2,3,\ldots$,

$$\varphi_n(x)=n\varphi(nx) \tag{14.13}$$

$$\psi_n(x)=\psi(x/n), \tag{14.14}$$

then we can construct sequences of test functions converging to $\tilde{f}$ as follows:

$$\lim_{n\to\infty} (\tilde{f} * \varphi_n)\psi_n = \tilde{f} \text{ in } D' \quad (\text{or } Z')$$

$$\lim_{n\to\infty} (\tilde{f}\psi_n) * \varphi_n = \tilde{f} \text{ in } D' \quad (\text{or } Z').$$

Given a functional $\tilde{f}$ in D' (or Z') and a convolute $\tilde{g}$ in D' (or Z'), we may define the convolution $\tilde{f}*\tilde{g}$ ($\equiv \tilde{g}*\tilde{f}$) as the function in D' (or Z') such that for each $\varphi \in D$ (or Z)

$$\langle \tilde{f}*\tilde{g}, \varphi\rangle = \langle \tilde{f}, \tilde{g}_{\mathrm{T}} * \varphi\rangle,$$

where $\tilde{g}_{\mathrm{T}}$ is the transpose of $\tilde{g}$: an alternative and equivalent definition may be based on the fact that with $\tilde{f}*\tilde{g}$ so defined we have

$$\lim_{n\to\infty} f_n * \tilde{g} = \tilde{f} * \tilde{g} \text{ in } D' \quad (\text{or } Z')$$

whenever $\{f_n\}$ is a sequence of test functions in D (or Z) converging to $\tilde{f}$ in D' (or Z').

If $\tilde{f}$ is in D' (or Z') and $\tilde{g}$ is a multiplier in D' (or Z') and $\tilde{h}$ is a convolute in D' (or Z') then

$$(\tilde{f}\tilde{g})' = \tilde{f}'\tilde{g} + \tilde{f}\tilde{g}'.$$

and $(\tilde{f}*\tilde{h})' = \tilde{f}'*\tilde{h} = \tilde{f}*\tilde{h}'$.

We now come to a pair of Fourier theorems on products and convolutions.

Theorem 14.12 Given $\tilde{f} \in D'$ with Fourier transform $\tilde{F} \in Z'$ and $\varphi \in D$ with transform $\Phi \in Z$, it follows that

$$\tilde{f}\varphi \leftrightarrow \tilde{F} * \Phi$$

and

$$\tilde{f} * \varphi \leftrightarrow \tilde{F}\Phi. \quad \square$$

Theorem 14.13 Suppose that $\tilde{f} \leftrightarrow \tilde{F}$ and $\tilde{g} \leftrightarrow \tilde{G}$, where $\tilde{f}$ and $\tilde{g}$ are each in class D': then (i) whenever $\tilde{g}$ is a convolute in D' we have

$$\tilde{f} * \tilde{g} \leftrightarrow \tilde{F}\tilde{G},$$

and whenever $\tilde{g}$ is a multiplier in D' we have

$$\tilde{f}\tilde{g} \leftrightarrow \tilde{F} * \tilde{G}. \quad \square$$

The extension of definitions of products and convolutions beyond the above stages is approached in different ways by different authors. One way is as follows, requiring joint restrictions on $\tilde{f}$ and $\tilde{g}$. Given $\tilde{f}$ and $\tilde{g}$ both in D' (or both in Z') we choose test functions $\varphi_n \in D$ (or Z) as in (14.13) and consider the limits as $n \to \infty$ of $\tilde{f}(\tilde{g} * \varphi_n)$ and of $\tilde{g}(\tilde{f} * \varphi_n)$: if these converge in D' (or Z') to a common limit functional which is independent of the particular choice of sequence φ_n, then we write the limit as $\tilde{f}\tilde{g}$ or equivalently as $\tilde{g}\tilde{f}$ and call it the product of $\tilde{f}$ and $\tilde{g}$. Likewise for a convolution product, given $\tilde{f}$ and $\tilde{g}$ both in D' (or both in Z') we choose test functions $\psi_n \in D$ (or Z) as in (14.14) and consider the limits as $n \to \infty$ of $\tilde{f} * (\tilde{g}\psi_n)$ and of $\tilde{g} * (\tilde{f}\psi_n)$: if these converge in D' (or Z') to a common limit functional which is independent of the particular choice of sequence ψ_n, then we write the limit as $\tilde{f} * \tilde{g}$ or equivalently as $\tilde{g} * \tilde{f}$ and call it the convolution of $\tilde{f}$ and $\tilde{g}$.

These definitions of $\tilde{f}\tilde{g}$ and $\tilde{f} * \tilde{g}$ are consistent with the definitions in previous sections. Also we find that if $\tilde{f}$ and $\tilde{g}$ are each in D' (or each in Z') and if the products $\tilde{f}\tilde{g}$ and $\tilde{f}\tilde{g}'$ exist, then $\tilde{f}'\tilde{g}$ will exist also and

$$(\tilde{f}\tilde{g})' = \tilde{f}'\tilde{g} + \tilde{f}\tilde{g}'.$$

Further if $\tilde{f}$ and $\tilde{g}$ are each in D' (or each in Z') and $\tilde{f} * \tilde{g}$ exists, then the following convolutions will exist and be equal:

$$(\tilde{f} * \tilde{g})' = \tilde{f}' * \tilde{g} = \tilde{f} * \tilde{g}'.$$

If f, g and $|f| * |g|$ are ordinary functions defining regular functionals all in D' (or all in Z') then $\tilde{f} * \tilde{g}$ will exist as a regular functional equal to $f * g$. However, the fact that f, g and fg are ordinary functions defining regular functionals all in D' (or all in Z') does not ensure that $\tilde{f}\tilde{g}$ exists: the functions in (13.25) and (13.26) provide counterexamples to prove this.

If $\tilde{f} \in D'$ is zero on $(-\infty, a)$ and $\tilde{g} \in D'$ is zero on $(a-\varepsilon, \infty)$ for real a and $\varepsilon > 0$, then $\tilde{f}\tilde{g}$ exists and equals zero. If $\tilde{f} \in D'$ equals zero on $(-\infty, a)$ and $\tilde{g} \in D'$ equals zero on $(-\infty, b)$, then $\tilde{f} * \tilde{g}$ exists in D' and equals zero on $(-\infty, c)$, where $c = a + b$ for any real a and b. Equation (13.27) gave an expression for $\delta^{(n)}f$ when $\tilde{f} \in S'$ satisfied certain conditions; this equation remains true if $\tilde{f}$ is in D' and satisfies the modified conditions obtained on replacing S' by D'. The results in this paragraph cannot necessarily be extended to functionals in Z' since local properties defined through (12.9) are not defined for all functionals in Z'.

We now arrive at the most general convolution and product theorems covered in this book.

Theorem 14.14 Suppose $\tilde{f} \leftrightarrow \tilde{F}$ and $\tilde{g} \leftrightarrow \tilde{G}$, where $\tilde{f}$ and $\tilde{g}$ are each in D': then (i) $\tilde{f}\tilde{g}$ will exist if, and only if, $\tilde{F} * \tilde{G}$ exists and in this case it will follow that

$$\tilde{f}\tilde{g} \leftrightarrow \tilde{F} * \tilde{G},$$

(ii) $\tilde{f} * \tilde{g}$ will exist if, and only if, $\tilde{F}\tilde{G}$ exists, and in this case it will follow that

$$\tilde{f} * \tilde{g} \leftrightarrow \tilde{F}\tilde{G}. \quad \square$$

15

Fourier series

15.1 Fourier coefficients of a periodic function

We say an ordinary function $f(x)$ is periodic if it is defined at a.a. x on $(-\infty, \infty)$ as a real or complex number and if there exists a positive number X such that for each $n=0, \pm 1, \pm 2, \ldots$

$$f(x+nX)=f(x) \tag{15.1}$$

whenever $f(x)$ is defined; we then say that f has period X. If f is of period X, then it is also of period $2X, 3X, \ldots$ so that the period of a function is not unique. In the following it is immaterial which period is chosen. Clearly a periodic function is uniquely determined on $(-\infty, \infty)$ by the values of $f(x)$ on an interval $[0, X]$.

If a periodic function f of period X is $L(0, X)$, and so integrable over every finite interval, then it follows that a set of complex numbers F_n, $n=0, \pm 1, \pm 2, \pm 3, \ldots$, may be defined by

$$F_n=\frac{1}{X}\int_{\lambda}^{\lambda+X} f(x)\exp(-2\pi inx/X)\mathrm{d}x; \tag{15.2}$$

these numbers are called the *Fourier series coefficients* of f, and their values are independent of the choice of (real) λ: moreover it can be shown that $\lim_{n\to\pm\infty} F_n=0$.

The following series, involving the Fourier series coefficients F_n of some function f of period X, is called the *Fourier series* belonging to f

$$\sum_{n=-\infty}^{+\infty} F_n \exp(2\pi inx/X). \tag{15.3}$$

Here and subsequently we take $\sum_{n=-\infty}^{+\infty}$ to mean the limit as $N\to\infty$ of $\sum_{n=-N}^{n=+N}$: it is important to note that the upper and lower limits run symmetrically to infinity. At any fixed value of x the Fourier series in (15.3) may or may not converge, depending on the choice of f and of x. The sum

s_N of a finite number of terms,

$$s_N(x) = \sum_{n=-N}^{+N} F_n \exp(2\pi i n x/X), \tag{15.4}$$

is called a *partial sum* of the Fourier series of f, and, at each N, $s_N(x)$ is a continuous function of x with period X. Many periodic functions possess a Fourier series that converges to the function itself at all x, whilst in other cases the Fourier series fails to converge to $f(x)$ at some or even all values of x. The central question of the theory of Fourier series is to determine in what sense and under what conditions $s_N(x)$ tends to $f(x)$ as $N \to \infty$.

Often the Fourier series is written in the equivalent form

$$\tfrac{1}{2}a_0 + \sum_{n=1}^{\infty} [a_n \cos(2\pi n x/X) + b_n \sin(2\pi n x/X)] \tag{15.5}$$

with

$$a_n = \frac{2}{X}\int_0^X f(x)\cos(2\pi n x/X)\mathrm{d}x \qquad (n = 0, 1, 2, \ldots) \tag{15.6}$$

$$b_n = \frac{2}{X}\int_0^X f(x)\sin(2\pi n x/X)\mathrm{d}x \qquad (n = 1, 2, 3, \ldots). \tag{15.7}$$

The coefficients a_n and b_n are called the *Fourier cosine and sine coefficients* and are related to the coefficients F_n as follows:

$$\begin{aligned} a_0 &= 2F_0 \\ a_n &= F_n + F_{-n} \qquad (n \geqslant 1) \\ b_n &= \mathrm{i}(F_n - F_{-n}) \qquad (n \geqslant 1). \end{aligned} \tag{15.8}$$

We will use the coefficients F_n, rather than a_n and b_n, because the F_n show a close relationship with the nomenclature of Fourier transforms.

The relation between a periodic locally integrable function and its Fourier coefficients is unique in the sense that two functions of equal period will have the same Fourier coefficients if, and only if, the functions are equal a.e. Of course two functions having different periods may share the same Fourier coefficients, but the functions will become essentially equal if the x variable is scaled.

15.2 The convergence of Fourier series

We start with a simple yet very useful set of conditions which ensure both the existence and the convergence of a Fourier series.

Theorem 15.1 Suppose that $f(x)$ is periodic, of period X, is defined and bounded on $[0, X]$ and that at least one of the following four conditions is satisfied: (i) f is piecewise monotonic on $[0, X]$, (ii) f has a finite number of maxima and minima on $[0, X]$ and a finite number of discontinuities on

$[0, X]$, (iii) f is of bounded variation on $[0, X]$, (iv) f is piecewise smooth on $[0, X]$: then it will follow that the Fourier series coefficients may be defined through (15.2), using proper Riemann integrals, and that the Fourier series (15.3) converges to $f(x)$ at a.a.x, to $f(x)$ at each point of continuity of f, and to the value $\frac{1}{2}[f(x^-)+f(x^+)]$ at all x. □

Despite the widespread applicability of the above result, the functions devised by du Bois-Reymond and by Fejér (section 5.5) warn us that the Fourier series of a periodic and everywhere continuous function may diverge at particular values of x or even at a non-denumerable and everywhere dense set of values of x. The following results, due to Carleson (1966) and Hunt (1968), show that the Fourier series of a continuous function cannot diverge everywhere.

Theorem 15.2 Suppose f is periodic, with period X, and suppose also that either (i) f is everywhere continuous, or, more generally, (ii) f is integrable over $[0, X]$ as a proper Riemann integral, or, more generally still, (iii) $f \in L^p(0, X)$ for some $1 < p \leqslant \infty$: then it follows that the Fourier series of f converges to $f(x)$ almost everywhere. □

In the above theorem the case $p=1$ is excluded, and the function devised by Kolmogoroff (see section 5.7) shows that a periodic function may be integrable (in the Lebesgue sense) over one period, and thus possess well defined Fourier coefficients, and yet have a Fourier series that diverges everywhere. A test for convergence at a particular point is as follows.

Theorem 15.3 Suppose $f(x)$ is periodic with period X and is $L(0, X)$: then the Fourier series of f will converge at $x=x_0$ if, and only if, x_0 is a Dirichlet point of f in which case the series converges to the Dirichlet value of f at $x=x_0$. □

Dirichlet points and Dirichlet values are defined in chapter 5, together with the tests of Jordan, of Dini, and of de la Vallée-Poussin: we recall that if $f(x)$ is differentiable at $x=x_0$ then this is a Dirichlet point with Dirichlet value $f(x_0)$, whilst if f is of bounded variation on $[x_0-\varepsilon, x_0+\varepsilon]$ then $x=x_0$ is a Dirichlet point of f with a Dirichlet value of $\frac{1}{2}[f(x_0^-)+f(x_0^+)]$.

Uniform convergence of a Fourier series over an interval may be secured as follows.

Theorem 15.5 Suppose f is periodic with period X and is $L(0, X)$, and suppose also that f is both continuous and of bounded variation on an interval $[a, b]$: then the partial sum $s_N(x)$ of the Fourier series of f, (15.4), will converge uniformly to $f(x)$ on $[c, d]$ as $N \to \infty$ whenever $a<c<d<b$. □

The partial sum s_N may be related to f in another way on introducing a function D_N often called the *Dirichlet kernel for Fourier series*. D_N is a continuous function defined through any of the following equivalent expressions, at all x, for $N=1,2,3,\ldots$:

$$D_N(x)=1+2\sum_{n=1}^{N}\cos nx=\sum_{n=-N}^{N}\exp inx$$
$$=\begin{cases}\dfrac{\sin(N+\frac{1}{2})x}{\sin(x/2)}, & x\neq\text{multiple of }2\pi\\ 2N+1, & x\text{ a multiple of }2\pi.\end{cases} \tag{15.9}$$

Theorem 15.6 Suppose f is of period X and is $L(0,X)$: then at all x, and for $N=1,2,3,4,\ldots,$

$$s_N(x)=\frac{1}{X}\int_{-X/2}^{X/2} f(x-u)D_N(2\pi u/X)\mathrm{d}u, \tag{15.10}$$

where $s_N(x)$ is the partial sum (15.4). □

15.3 Summability of Fourier series

Consider a series $\sum_{n=1}^{\infty} a_n$ of real or complex numbers. The series is said to be *Cesàro summable* if

$$\lim_{N\to\infty}\left[\sum_{n=1}^{N} a_n\left(1-\frac{n}{N}\right)\right] \tag{15.11}$$

converges, and the limit value is then called the *Cesàro sum* of the series. Likewise if $\sum_{n=1}^{\infty} a_n \exp(-n/\lambda)$ converges for each $\lambda>0$ and if

$$\lim_{\lambda\to\infty}\left[\sum_{n=1}^{\infty} a_n\exp(-n/\lambda)\right] \tag{15.12}$$

converges, then the series is said to be *Abel summable* and the limit is called the *Abel sum* of the series; the index λ may run continuously to infinity, or through integer values.

If the series $\sum_{n=1}^{\infty} a_n$ is convergent in the ordinary sense then it will certainly be Cesàro summable and Abel summable, and all three limits will be the same. However, a series that does not converge in the ordinary sense (i.e. it diverges) may yet be summable in the Cesàro and/or Abel senses. The factors $[1-(n/N)]$ and $\exp(-n/\lambda)$ play a role analogous to that of the convergence factors appearing in integrands in chapter 8. For example, the divergent series $1-1+1-1+1-\cdots$ is Cesàro and Abel summable to the value $\frac{1}{2}$ in each case. The divergent series $1-2+3-4+5-\cdots$ is, however, not Cesàro summable, though it is Abel summable to the value $\frac{1}{4}$. Sometimes summability can imply ordinary convergence, as in the following cases proved by Hardy and Littlewood for Cesàro and Abel sums, respectively. If a series $\sum_{n=1}^{\infty} a_n$ has $a_n = O(1/n)$ then Cesàro summability will imply ordinary convergence to the same limit, and likewise Abel summability will imply convergence to the same limit. The condition $a_n = O(1/n)$ means, more explicitly, that there exists an $A > 0$ such that $|a_n| \leqslant A/n$ for all $n \geqslant 1$.

Clearly a Fourier series might be summable when it is not convergent in the ordinary sense, and the use of summability allows the constraints on $f(x)$ to be relaxed.

Theorem 15.7 Suppose f is periodic with period X and that $f \in L(0, X)$: then Fourier coefficients F_n may be defined through (15.2) and each of

$$\lim_{N \to \infty} \sum_{n=-N}^{+N} F_n \left(1 - \frac{|n|}{N}\right) \exp(2\pi i n x/X) \tag{15.13}$$

and

$$\lim_{\lambda \to \infty} \sum_{n=-\infty}^{+\infty} F_n \exp\left(-\frac{|n|}{\lambda}\right) \exp(2\pi i n x/X) \tag{15.14}$$

will converge in the following ways: (i) to $f(x)$ a.e., (ii) to $f(x)$ at each point of continuity of f, (iii) uniformly to $f(x)$ on an interval $[a, b]$ whenever f is continuous on $[a-\varepsilon, b+\varepsilon]$ for some $\varepsilon > 0$, (iv) to $\frac{1}{2}[f(x^-) + f(x^+)]$ at each point where $f(x^-)$ and $f(x^+)$ exist, (v) to the Lebesgue value of $f(x)$ at each Lebesgue point of $f(x)$, (vi) as a limit in the mean, index 1, over $[0, X]$. □

The meanings of Lebesgue points and Lebesgue values are described in section 5.2.

The summability methods of Cesàro and Abel are not the only ones that can be used with Fourier series, and other convergence factors may be generated as follows.

Theorem 15.8 Suppose that f is periodic with period X and that $f \in L(0, X)$, and suppose that K is an everywhere continuous function defined on $(-\infty, \infty)$ satisfying conditions (C1)–(C4) of theorem 8.10: then it will follow that for $\lambda > 0$

$$\lim_{\lambda \to \infty} \sum_{n=-\infty}^{+\infty} F_n K(n/\lambda) \exp(2\pi i n x/X) \tag{15.15}$$

will converge as in (i)–(vi) of theorem 15.7, where the F_n are the Fourier series coefficients defined by (15.2). □

15.4 Mean convergence of Fourier series

A locally integrable periodic function f necessarily has Fourier coefficients F_n that tend to zero as $n \to \pm\infty$. However, a two-way sequence of numbers, $\{F_n\}_{n=-\infty}^{\infty}$, such that $\lim_{n \to \pm\infty} F_n = 0$ will not necessarily consist of the Fourier coefficients of any locally integrable periodic function: indeed even if

$$\sum_{n=-\infty}^{\infty} F_n \exp(2\pi i n x/X) \tag{15.16}$$

converges at a.a.x to define some function $f(x)$, it does not necessarily follow that the F_n will be the Fourier coefficients of f, (15.2), since f may not be locally integrable. It is true, however, that if (15.16) converges at all x to define a locally integrable function f, then the F_n will be the Fourier coefficients of f.

In fact no simple test on a two-way sequence of numbers is known that is both necessary and sufficient to ensure that they are the Fourier coefficients of some periodic $f \in L_{\text{LOC}}$. Various simple sufficient conditions are known, and convexity provides one such test.

Theorem 15.9 Let $\{F_n\}_{n=-\infty}^{\infty}$ be a sequence of non-negative real numbers such that (i) $F_n = F_{-n}$ at each n, (ii) $\lim_{n\to\infty} F_n = 0$, and (iii) at each $n \geqslant 0$

$$F_n - F_{n+1} \geqslant F_{n+1} - F_{n+2}:$$

then it will follow that the F_n are the Fourier coefficients of some locally integrable periodic function that is nowhere negative. □

For example, the set $F_0 = 2$, $F_n = F_{-n} = |n|^{-1}$ ($|n| \geqslant 1$) satisfy the conditions of the above theorem; it is true that the corresponding periodic function $f(x)$ diverges around $x = 0$, but the singularity is integrable. Likewise the two-way sequence $F_n = [\ln(2 + |n|)]^{-1}$, $n = 0, \pm 1, \pm 2, \ldots$, satisfies the conditions of theorem 15.9. However, the set $F_0 = (\ln 2)^{-1}$, $F_n = \operatorname{sgn} n/[\ln(2 + |n|)]$, $n = \pm 1, \pm 2, \ldots$, fails this test: in fact the trigonometric series (15.16) based on this sequence converges at *all* x to define a periodic function $f(x)$, but $f(x)$ is not integrable over one period due to infinite discontinuities at $x = 0, \pm X, \pm 2X, \ldots$.

Another type of condition is as follows.

Theorem 15.10 Let $\{F_n\}_{n=-\infty}^{\infty}$ be a sequence of numbers such that, for some p satisfying $1 \leqslant p \leqslant 2$, $\sum_{n=-\infty}^{\infty} |F_n|^p$ converges: then the F_n will be the Fourier coefficients of some periodic locally integrable function f; moreover when $1 < p \leqslant 2$ then $f \in L^q_{\text{LOC}}$ ($p^{-1} + q^{-1} = 1$), whilst when $p = 1$ then the F_n are the Fourier coefficients of a continuous periodic function. □

There is a partial converse to the above.

Theorem 15.11 Let f be a periodic function that is L^p_{LOC} for some p satisfying $1 \leqslant p \leqslant 2$: then it follows that $f \in L^1_{\text{LOC}}$ and the Fourier coefficients F_n, (15.2), are defined, and $\lim_{n\to\infty} F_n = 0$; moreover when $1 < p \leqslant 2$ then $\sum_{n=-\infty}^{\infty} |F_n|^q$ will converge, where $p^{-1} + q^{-1} = 1$. □

As might be expected, pth power integrability is accompanied by mean convergence of the Fourier series, as follows.

Theorem 15.12 Let f be a periodic function of period X that is $L^p(0, X)$ for some p satisfying $1 < p < \infty$: then also $f \in L^1(0, X)$ and Fourier coefficients are defined, and for each q such that $1 \leqslant q \leqslant p$,

$$\lim_{N \to \infty} \int_0^X |s_N(x) - f(x)|^q \mathrm{d}x = 0, \tag{15.17}$$

where $s_N(x)$ is the partial sum (15.4) of the Fourier series. □

The case $p = 1$, not covered in the above theorem on account of the Kolmogoroff example, (section 5.7) is covered by the summability methods in theorems (15.7) and (15.8). The case $p = 1$ is also covered by the following.

Theorem 15.13 Suppose f is periodic with period X, is $L(0, X)$, and is of bounded variation on $[0, X]$: then it follows that Fourier coefficients F_n may be defined, and that $F_n = O(1/|n|)$ for all n, and that

$$\lim_{N \to \infty} \int_0^X |s_N(x) - f(x)| \mathrm{d}x = 0$$

where $s_N(x)$ is the partial sum, (15.4), of the Fourier series. □

The case $p = 2$ is special in that theorems 15.10 and 15.11 become strict converses of each other. In fact most of the cases met in elementary applications fall in this category, and we now list some especially important results.

Theorem 15.14 Each periodic function that is L^2 over one period will have Fourier coefficients such that

$$\sum_{n=-\infty}^{\infty} |F_n|^2 < \infty \tag{15.18}$$

converges, and conversely any two-way sequence of numbers satisfying (15.18) will be the Fourier coefficients of some periodic function that is L^2 over one period. □

In the first part of theorem 15.14 one may replace the condition of quadratic Lebesgue integrability over one period by the condition that the function is quadratically Riemann integrable over one period, as a proper or improper integral: however, in the second part of the theorem the convergence of (15.18) is *not* sufficient to ensure that the F_n are the coefficients of some locally quadratically Riemann integrable function. This is an example of the way in which the more powerful Lebesgue method leads to simpler results than the Riemann method.

There is a sense in which a best fit of a trigonometric polynomial (with a fixed finite number of terms) will yield some of the Fourier coefficients, as follows.

Theorem 15.15 Let f be a periodic function, of period X, such that $f \in L^2(0, X)$, and consider a set of numbers A_n, $n=0, \pm 1, \pm 2, \ldots, \pm N$, together with the sum

$$S_N = \sum_{n=-N}^{N} A_n \exp(2\pi i n x/X):$$

then it will follow that

$$\int_0^X |S_N(x) - f(x)|^2 \mathrm{d}x \geqslant \int_0^X |s_N(x) - f(x)|^2 \mathrm{d}x \tag{15.19}$$

where $s_N(x)$ is the partial sum, (15.4), of the Fourier series of f; moreover the equality sign in (15.19) will hold if, and only if, each A_n is equal to the corresponding Fourier coefficient for $|n| \leqslant N$. □

15.5 Sampling theorems

The sampling theorem in section 8.5 is related to several others. If $\varphi \leftrightarrow \Phi$ is a Fourier pair of good functions and X is a fixed positive number, then

$$X \sum_{n=-\infty}^{\infty} \varphi(x - nX) = \sum_{n=-\infty}^{\infty} \Phi(n/X) \exp(2\pi i n x/X) \tag{15.20}$$

where the left hand series converges pointwise at each x on $(-\infty, \infty)$, whilst the right hand series converges uniformly on $x \in (-\infty, \infty)$. In other words, repeated additions of shifted versions of the function φ yields a periodic function, say $g(x)$, equal to the left side of (15.20), whose Fourier coefficients G_n are equal to sampled values of Φ,

$$G_n = \Phi(n/X) \qquad (n=0, \pm 1, \pm 2, \ldots). \tag{15.21}$$

On putting $x=0$ in (15.20) we obtain *Poisson's formula*,

$$X \sum_{n=-\infty}^{\infty} \varphi(nX) = \sum_{n=-\infty}^{\infty} \Phi(n/X). \tag{15.22}$$

If in the above discussion $\varphi(x)$ is equal to zero for $|x| \geqslant X/2$, then the graph of $g(x)$ simply consists of a set of equally spaced replicas of $\varphi(x)$, centred on the points $x = nX, n=0, \pm 1, \pm 2, \ldots$. If $\varphi(x)$ does not equal zero for $|x| > x$, then 'overlaps' occur in the left hand summation in (15.20): this overlap is sometimes referred to as *aliasing*. Aliasing occurs if the sample points in theorem 8.29 are separated by more than the Nyquist interval $1/2a$.

The above equations can hold for functions other than good functions, and the following theorems describe sufficient conditions.

Theorem 15.16 Suppose that $f(x)\leftrightarrow F(y)$ is a Fourier pair and that at least one of these functions is both $L(-\infty,\infty)$ and also of bounded variation on $(-\infty,\infty)$: then f and F are a.e. equal, respectively, to functions f_D and F_D such that at all x

$$f_D(x)=\tfrac{1}{2}[f_D(x^-)+f_D(x^+)]$$
$$F_D(x)=\tfrac{1}{2}[F_D(x^-)+F_D(x^+)],$$

and it will follow that, for any fixed $X>0$,

$$\text{(i)}\quad X\sum_{n=-\infty}^{\infty} f_D(x-nX)=\sum_{n=-\infty}^{+\infty} F_D(n/X)\exp(2\pi inx/X) \tag{15.23}$$

where each series converges pointwise on $x\in(-\infty,\infty)$, and

(ii) the left side of (15.23) defines a function, say $g(x)$, of period X that is $L(0,X)$ and has Fourier coefficients G_n such that

$$G_n=F_D(n/X)\qquad (n=0,\pm 1,\pm 2,\ldots), \tag{15.24}$$

$$\text{(iii)}\quad X\sum_{n=-\infty}^{\infty} f_D(nX)=\sum_{n=-\infty}^{+\infty} F_D(n/X), \tag{15.25}$$

(iv) when F is $L(-\infty,\infty)$ and of bounded variation on $(-\infty,\infty)$ then the Fourier series of g converges uniformly on $(-\infty,\infty)$,

(v) when f is $L(-\infty,\infty)$ and of bounded variation on $(-\infty,\infty)$, then g is of bounded variation on $[0,X]$. □

If the condition of bounded variation is dropped, some of the above results still remain, as follows.

Theorem 15.17 Suppose $f\in L(-\infty,\infty)$ and let F be the everywhere continuous Fourier transform of f: then for any fixed $X>0$ it follows that

$$X\sum_{n=-\infty}^{+\infty} f(x-nX)$$

converges a.e. to define a function $g(x)$ that is $L(0,X)$ and of period X, and the Fourier coefficients G_n of g will be

$$G_n=F(n/X)\qquad n=0,\pm 1,\pm 2,\ldots;$$

however, the Fourier series of g will not necessarily converge at any point, and the Poisson sum formula will not necessarily be valid. □

The Fourier pair (8.5) can be used as an example of the use of theorem 15.16. For instance, for some positive integer N, consider $f\leftrightarrow F$, where

$$F(y)=F_D(y)=\begin{cases}1, & -(N+\tfrac{1}{2})<y<(N+\tfrac{1}{2})\\ \tfrac{1}{2}, & y=\pm(N+\tfrac{1}{2})\\ 0, & |y|>(N+\tfrac{1}{2})\end{cases}$$

so that

$$f(x)=f_{\mathrm{D}}(x)=\begin{cases}\dfrac{\sin[2\pi(N+\frac{1}{2})x]}{\pi x}, & x\neq 0\\ 2N+1, & x=0.\end{cases}$$

Equation (15.23), with $X=1$, now leads to a relation between the Dirichlet kernel $\sin x/\pi x$ used with Fourier transforms, and the Dirichlet kernel $D_N(x)$ used with Fourier series,

$$\sum_{n=-\infty}^{+\infty}\frac{\sin[2\pi(N+\frac{1}{2})(x-n)]}{\pi(x-n)}=1+2\sum_{n=1}^{N}\cos(2\pi nx)$$
$$=D_N(2\pi x).$$

15.6 Differentiation and integration of Fourier series

If a periodic function, $f(x)$, of period X, has an everywhere continuous derivative $f'(x)$ then the derivative will have a Fourier series that can be obtained from that of $f(x)$ by differentiating the series term by term: this is equivalent to saying that the operation of differentiating $f(x)$ has the effect of multiplying the nth Fourier coefficient by $2\pi in/X$. This rule can remain valid even when f is not everywhere differentiable, and we now provide some sets of conditions for its validity. We can call them classical differentiation theorems, to distinguish them from the more powerful results which are possible when generalized functions are used.

Theorem 15.18 Suppose functions f and g are each $L(0, X)$ and of period X, and let the Fourier coefficients, (15.2), be, respectively, $\{F_n\}$ and $\{G_n\}$: then the Fourier series of g will be obtained by term by term differentiation of the Fourier series of f, and

$$G_n=(2\pi in/X)F_n \qquad (n=0, \pm 1, \pm 2, \ldots) \tag{15.26}$$

if, and only if, f is equal a.e. on $(-\infty, \infty)$ to an indefinite integral of g. □

When the conditions of this theorem are met it will mean that f will equal, a.e. on $(-\infty, \infty)$, an absolutely continuous function, and that g will equal a.e. the derivative of f. Clearly the conditions of theorem 15.18 will *not* be met if $f(x)$ possesses any form of discontinuity other than a removable one. Thus, for example, if f has period X and is specified by $f(x)=x$ on $(-X/2, X/2]$, term by term, differentiation of its Fourier series will *not* give the Fourier series of $g(x)=1$ (all x); this is because f has step discontinuities at $x=\pm X/2$.

The various conditions for ensuring absolute continuity, section 3.6, all provide special forms of theorem 15.18. One simple one is as follows.

Theorem 15.19 Consider a function f that is continuous on $(-\infty, \infty)$, has period X, and has a derivative f' that is continuous except at a finite number of points on $[0, X]$, and suppose also that either (i) $|f'(x)|$ nowhere has a value greater than some fixed number $M>0$, or (ii) f' is $L(0, X)$: then f and f' will possess Fourier coefficients, the Fourier series of f' will be obtained from that of f by term by term differentiation, and the Fourier coefficients G_n of f' will be related to the coefficients F_n of f by (15.26) for $n=0, \pm 1, \pm 2, \ldots$. □

As examples consider four functions, each of period X, having the following values on $[-X/2, X/2]$:

$$f_1(x)=|x|, \qquad -X/2 \leqslant x \leqslant X/2$$

$$f_2(x)=|x|^{1/2}, \qquad -X/2 \leqslant x \leqslant X/2$$

$$f_3(x)=\begin{cases} x^2\cos(1/x), & 0<|x| \leqslant X/2 \\ 0, & x=0 \end{cases}$$

$$f_4(x)=\begin{cases} x^2\cos(1/x^2), & 0<|x| \leqslant X/2 \\ 0, & x=0. \end{cases}$$

The functions f_1, f_2 and f_3 each satisfy the conditions of theorem 15.19. Although f_4 possesses a derivative at a.a.x, it does not satisfy the conditions of theorem 15.18 or theorem 15.19 because the derivative is not integrable; this case can, however, be handled using the concepts of generalized functions (chapter 16).

Integration of Fourier series presents less difficulties than differentiation, in that term by term integration of the Fourier series of a periodic function always converges pointwise to an integral of the function, as follows.

Theorem 15.20 Suppose $f \in L(0, X)$ has period X and Fourier coefficients F_n: then for each real a and each real x

$$\int_a^x f(u)\mathrm{d}u = \sum_{n=-\infty}^{\infty} \left[\int_a^x F_n \exp(2\pi i n u/X)\mathrm{d}u\right]. \quad \Box \qquad (15.27)$$

If we fix the value of a then both sides of (15.27) define pointwise a function, say $g(x)$, which is an indefinite integral of f. $g(x)$ will be periodic if, and only if, $F_0=0$. When the primitive $g(x)$ is periodic it will have period X and its Fourier coefficients will be

$$G_n=\frac{F_n}{2\pi i n} \qquad (n=\pm 1, \pm 2, \ldots),$$

whilst the value of G_0 will depend on the value of a according to

$$G_0=-\sum_{n=1}^{\infty} [G_n \exp(2\pi i n a/X) - G_{-n} \exp(-2\pi i n a/X)].$$

15.7 Products and convolutions

Just as the Fourier transform of the product or convolution of two functions is equal to the convolution or product, respectively, of the individual transforms, so there are analogous results for periodic functions and their sets of Fourier coefficients. There are also hybrid results in which periodic functions are multiplied or convoluted with transformable functions that are $L^p(-\infty, \infty)$.

We start with the convolution between two periodic functions, which is one of the simplest cases.

Theorem 15.21 Suppose f and g are each of period X and $L(0, X)$, and let $\{F_n\}$ and $\{G_n\}$ be their Fourier coefficients: then (i) at a.a.x we may define a function $k(x)$, called the T-convolution of f and g, by

$$k(x) = \frac{1}{X}\int_0^X f(x-x')g(x')\,dx', \tag{15.28}$$

(ii) the convolution k will be of period X, will be $L(0, X)$ and will have Fourier coefficients K_n given by

$$K_n = F_n G_n \qquad (n=0, \pm 1, \pm 2, \ldots). \quad \square \tag{15.29}$$

Note that the convolution $k(x)$ will be unaltered if the integration limits in (15.28) are replaced by $\int_\lambda^{\lambda+X}$ for arbitrary real λ. Quite commonly the T-convolution is defined without the factor $(1/X)$ in front of the integral in (15.28), but we will adopt (15.28) as it stands on account of the simplicity of (15.29).

The condition $L^p(0, X)$ for some $p > 1$ implies also the condition $L^r(0, X)$ whenever $1 \leqslant r \leqslant p$, and this would constitute a stiffening in the conditions on a periodic function if inserted into theorem 15.21. In particular if the functions f and g in theorem 15.21 are each $L^2(0, X)$ then the convolution $k(x)$ will be everywhere continuous and will have a Fourier series that converges uniformly to $k(x)$ on $(-\infty, \infty)$. It is further true that if, with f and g as in theorem 15.21 we have $f \in L^p(0, X)$ and $g \in L^r(0, X)$, where p and r satisfy $p \in [1, \infty]$, $q \in [1, \infty]$, $p^{-1} + q^{-1} \geqslant 1$, then it will follow that the convolution $k(x)$, (15.28), will be $L^t(0, X)$, where $p^{-1} + r^{-1} = 1 + t^{-1}$. throughout we regard ∞^{-1} and zero as interchangeable.

A corresponding theorem for the product of two periodic functions is more complicated because conditions have to be applied to ensure that the product is integrable, and also because the idea of convoluting two sets of Fourier coefficients needs care (especially when they only decay slowly as $n \to \infty$). The following gives some useful sets of conditions.

Theorem 15.22 Suppose that f and g are each of period X and $L(0, X)$, and that their Fourier coefficients are $\{F_n\}$ and $\{G_n\}$, and suppose also that at least one of the following conditions (i)–(iii) is satisfied: (i) $f \in L^2(0, X)$ and

$g \in L^2(0, X)$, (ii) f and/or g is of bounded variation on $[0, X]$, (iii) $f \in L^p(0, X)$ and $g \in L^q(0, X)$ for some p and q satisfying $1 \leqslant p \leqslant \infty$, $1 \leqslant q \leqslant \infty$, $p^{-1} + q^{-1} = 1$ (with $\infty^{-1} \equiv 0$ as usual): then it will follow that the product $k = fg$ is $L(0, X)$ and of period X and will have Fourier coefficients, say $\{K_n\}$; moreover it will follow that at each n $(=0, \pm 1, \pm 2, \ldots)$

$$K_n = \sum_{m=-\infty}^{\infty} F_m G_{n-m} = \sum_{m=-\infty}^{\infty} F_{n-m} G_m \tag{15.30}$$

where (15.30) converges absolutely in case (i), in the ordinary sense in case (ii), and as a Cesàro sum in case (iii), which means that

$$K_n = \lim_{M \to \infty} \sum_{m=-M}^{M} F_m G_{n-m} \left(1 - \frac{|n|}{M}\right). \quad \square \tag{15.31}$$

The summations in (15.30) and (15.31) can be regarded as defining convolutions between the sets $\{F_n\}$ and $\{G_n\}$.

The Parseval equation for Fourier series emerges as a special case of the above (on evaluating K_0) but is worth dealing with separately.

Theorem 15.23 Suppose f and g are of period X and $L(0, X)$ and that at least one of the conditions (i), (ii) and (iii) of theorem 15.22 is satisfied: then it will follow that f and g will have Fourier coefficients, say $\{F_n\}$ and $\{G_n\}$, and that

$$\frac{1}{X} \int_0^X f g^* = \sum_{n=-\infty}^{\infty} F_n G_n^* \tag{15.32}$$

where in case (i) the series is absolutely convergent, in case (ii) the series is convergent in the ordinary sense, whilst in case (iii) the series is Cesàro summable so that,

$$\frac{1}{X} \int_0^X f g^* = \lim_{N \to \infty} \sum_{n=-N}^{N} F_n G_n^* \left(1 - \frac{|n|}{N}\right). \quad \square \tag{15.33}$$

Equations (15.32) and (15.33) are versions of Parseval's formula for Fourier series. Each can be modified to an equivalent form such as the following absolute, ordinary, or Cesàro sum,

$$\frac{1}{X} \int_0^X f g = \sum_{n=-\infty}^{\infty} F_n G_{-n}. \tag{15.34}$$

When $f = g$ the conditions in theorem 15.22 imply that $f \in L^2(0, X)$.

We thus arrive at

Theorem 15.14 Suppose f has period X: then if, and only if, $f \in L^2(0, X)$ will it follow that

$$\frac{1}{X}\int_0^X |f|^2 = \sum_{n=-\infty}^{\infty} |F_n|^2, \qquad (15.35)$$

where the F_n are the Fourier coefficients of f, (15.2). □

The Parseval equation, (15.35), can also be regarded as a limiting case of the following inequalities.

Theorem 15.25 Suppose f has period X and is $L(0, X)$, and let the Fourier coefficients be F_n: then (i) whenever $f \in L^p(0, X)$ for some $p \in (1, 2)$ it follows that

$$\left\{\frac{1}{X}\int_0^X |f(x)|^p \mathrm{d}x\right\}^{1/p} \geqslant \left\{\sum_{n=-\infty}^{\infty} |F_n|^q\right\}^{1/q}, \qquad (15.36)$$

and (ii) whenever $\sum |F_n|^p < \infty$ for some $p \in (1, 2)$ it follows that

$$\left\{\frac{1}{X}\int_0^X |f(x)|^q \mathrm{d}x\right\}^{1/q} \leqslant \left\{\sum_{n=-\infty}^{\infty} |F_n|^p\right\}^{1/p}, \qquad (15.37)$$

where in both cases $p^{-1} + q^{-1} = 1$. □

The smoothing or truncation of a periodic function can be represented by convolution or multiplication with some function that is $L(-\infty, \infty)$. The following theorems are relevant to this case.

Theorem 15.25 Suppose f has period X and is $L(0, X)$ with Fourier coefficients $\{F_n\}$ and that $g \in L(-\infty, \infty)$ has the continuous Fourier transform G: then we may define a convolution $k(x)$ at a.a.x by

$$k(x) = \int_{-\infty}^{\infty} f(x')g(x-x')\mathrm{d}x', \qquad (15.38)$$

and k will be L_{LOC} and of period X with Fourier coefficients K_n given by

$$K_n = F_n G(n/X) \qquad (n = 0, \pm 1, \pm 2, \ldots). \; \square \qquad (15.39)$$

In the above theorem a corresponding product theorem will not necessarily be possible because fg will not necessarily be L_{LOC}. However, an additional condition allows further results.

Theorem 15.26 Suppose f and g satisfy the conditions of theorem 15.25 and that in addition either (*a*) g is of bounded variation on $(-\infty, \infty)$, or (*b*) $g(x)$ and $G(x)$ are each $O[1/(1+|x|)^s]$ for some $s > 1$: then (i) (15.38) and (15.39) will be applicable, the convolution $k(x)$ being defined at all x, (ii) the product $h = fg$ will be $L(-\infty, \infty)$ and will possess a continuous Fourier transform $H(y)$ given at each y (and indeed uniformly on $(-\infty, \infty)$) by

$$H(y) = \sum_{n=-\infty}^{+\infty} F_n G(y - n/X),$$

and (iii) the following Parseval formula will be valid:

$$\int_{-\infty}^{+\infty} fg^* = \sum_{n=-\infty}^{+\infty} F_n G^*(n/X). \quad \square$$

16

Generalized Fourier series

16.1 Introduction

A periodic locally integrable function defines a regular functional in S' which will therefore possess a Fourier transform in S'. It turns out that this Fourier transform consists of equally spaced delta functions which are weighted according to the Fourier coefficients of the function. The use of generalized functions thus allows the subject of Fourier series to be considered as a special case of Fourier transformation. It achieves more than this, however, since we move from the analysis of periodic functions to that of periodic functionals.

A functional $\tilde{f}$ in any one of the classes S', D' or Z' is said to be periodic if there exists a positive number X such that for each $n=0, \pm 1, \pm 2, \ldots$,

$$\tilde{f}(x+nX)=\tilde{f}(x):$$

the functional is then said to have period X. As with ordinary functions a functional having period X will also have period $2X, 3X, 4X, \ldots$, and in what follows it is immaterial which period is chosen. As a simple example it can be shown that for any fixed $X>0$,

$$\lim_{N\to\infty} \sum_{n=-N}^{N} \tilde{\delta}(x-nX) \tag{16.1}$$

converges in S' to a functional of period X: we write such a limit as $\sum_{n=-\infty}^{\infty} \tilde{\delta}(x-nX)$. Here and subsequently $\sum_{n=-\infty}^{\infty}$ is understood as the symmetrical type of limit just used in (16.1).

It can be shown that a periodic functional in D' will necessarily also be a periodic functional in S', but a periodic functional in Z' will not necessarily be either in D' or in S'. We therefore need to consider only periodic functionals in Z' or S', the latter being a more restricted class than the former.

16.2 Generalized Fourier coefficients

The following theorems set up the fundamentals of generalized Fourier series.

Theorem 16.1 Let $\tilde{f} \in Z'$ have period X: then it follows that the Fourier transform $\tilde{F}$ of $\tilde{f}$ can be written in the form

$$\tilde{F}(y) = \sum_{n=-\infty}^{\infty} F_n \tilde{\delta}(y - n/X), \tag{16.2}$$

where the F_n, $n = 0, \pm 1, \pm 2, \ldots$, are complex numbers and the summation converges in D'; moreover if in addition $\tilde{f} \in S'$ then it will follow that there will exist an $A > 0$ and an $M > 0$ such that for each n

$$|F_n| \leqslant A|n|^M, \tag{16.3}$$

and the summation in (16.2) will converge in S'. □

As is natural we define the numbers F_n as the *generalized Fourier coefficients* of $\tilde{f}$, or simply as the Fourier coefficients of $\tilde{f}$ when no ambiguity will arise. When $\tilde{f}$ is regular and periodic the generalized Fourier coefficients are identical to the ordinary Fourier coefficients, (15.2).

The generalized form of Fourier's theorem for series is now as follows.

Theorem 16.2 Suppose $\tilde{f}$ is in class Z', is of period X, and has Fourier coefficients $\{F_n\}$: then it follows that

$$\tilde{f}(x) = \sum_{n=-\infty}^{\infty} F_n \exp(2\pi i n x/X) \tag{16.4}$$

where the series converges in Z'; when in addition $\tilde{f} \in S'$ then the series converges in S'. □

Naturally we define the series in (16.4) as the *generalized Fourier series of* $\tilde{f}$, or simply as the Fourier series of $\tilde{f}$ when no ambiguity will arise.

Generalized Fourier series are unique in the sense that two periodic functionals of equal period will be equal if, and only if, their Fourier coefficients are all equal, and conversely.

The question of whether a two-way sequence of numbers will be the Fourier coefficients of some periodic functional is easier to answer in generalized analysis than in the classical context.

Theorem 16.3 Let $\{F_n\}$, $n = 0, \pm 1, \pm 2, \ldots$, be a two-way sequence of real or complex numbers: then for each $X > 0$ the series

$$\sum_{n=-\infty}^{\infty} F_n \exp(2\pi i n x/X) \tag{16.5}$$

will converge in Z' to a functional, say $\tilde{f}$, in Z' of period X whose Fourier coefficients are the $\{F_n\}$; if moreover the F_n satisfy

$$|F_n| \leqslant A|n|^M$$

for some $A>0$ and $M>0$ then $\tilde{f}$ will be in S' and the series in (16.5) will converge in S'. □

For example, the numbers $F_n = 2\pi n$ $(n=0, \pm 1, \pm 2, \ldots)$ must be Fourier coefficients, and in fact the corresponding functional of period X is

$$\tilde{f} = -\mathrm{i}X^2 \sum_{n=-\infty}^{\infty} \delta'(x-nX) \qquad (\text{in } S'). \tag{16.6}$$

16.3 The Fourier formulae

As yet, we lack a formula for obtaining the Fourier coefficients of some periodic functional; we cannot use

$$\frac{1}{X}\int_0^X \tilde{f} \exp(-2\pi \mathrm{i}nx/X)\mathrm{d}x$$

because meaning has not been given to such an integral when $\tilde{f}$ is a generalized function. Entities called unitary functions provide the ingenious way out of this difficulty. An ordinary function $h(x)$ is said to be *unitary with parameter* $X(>0)$ if at all x

$$\sum_{n=-\infty}^{+\infty} h(x-nX) = 1, \tag{16.7}$$

using ordinary convergence. The reader will readily verify that the 'triangular' function

$$h(x) = \begin{cases} 1-|x|, & -1<x<1 \\ 0, & |x| \geqslant 1 \end{cases} \tag{16.8}$$

is unitary with parameter unity. A function is said to be *unitary in S (or D, or Z)* with parameter X if the function is in class S (or D, or Z) and is unitary with parameter X. For example, the function

$$h(x) = \frac{\displaystyle\int_{|x|}^1 \exp[-(u-u^2)^{-1}]\mathrm{d}u}{\displaystyle\int_0^1 \exp[-(u-u^2)^{-1}]\mathrm{d}u}, \qquad -1<x<1$$

$$h(x) = 0, \qquad |x| \geqslant 1$$

is unitary in D and in S (but not in Z) with parameter $X=1$. A unitary function in Z with parameter X may be obtained by Fourier transforming a function ρ in D which has zero value outside of $[-1/X, 1/X]$ and has $\rho(0)=X$. Thus $h(x)$ defined as follows for all x on $(-\infty, \infty)$ is a unitary function in Z', with parameter X whenever $0<X\leqslant 1$,

$$h(x)=\int_{-1}^{+1}\rho(y)\exp(2\pi ixy)\mathrm{d}y,$$

where

$$\rho(y)=X\exp\left(\frac{-y^2}{1-y^2}\right)\qquad -1<y<1.$$

A unitary function in Z is necessarily unitary in S, but not conversely.

If $f\in L(0,X)$ has period X then it will follow that whenever $h(x)$ is a continuous unitary function with parameter X, then

$$\int_0^X f(x)\mathrm{d}x=\int_{-\infty}^{\infty}f(x)h(x)\mathrm{d}x. \tag{16.9}$$

The reader will readily verify this if h is the triangular function, (16.8). With this fact in mind the following generalized version of Fourier's formula will perhaps seem less surprising than otherwise.

Theorem 16.4 Let $\tilde{f}$ be a functional in Z' with period X: then the Fourier coefficients $\{F_n\}$ are given by

$$F_n=\frac{1}{X}\langle\tilde{f}(x)\exp(-2\pi inx/X),\,h(x)\rangle, \tag{16.10}$$

where $h(x)$ is any unitary function in Z with parameter X; in the special case that $\tilde{f}\in S'$ then $h(x)$ may be a unitary function in S' with parameter X. □

16.4 Differentiation, repetition and sampling

Many of the theorems in chapter 15 have their generalized counterparts, which run rather more smoothly on the whole. For instance the differentiation and integration theorems.

Theorem 16.5 Let $\tilde{f}$ be a functional in Z' or in S' with period X having Fourier coefficients $\{F_n\}$: then it follows that the derivative say $\tilde{g}=\tilde{f}'$ will also have period X and will have Fourier coefficients

$$G_n=\left(\frac{2\pi in}{X}\right)F_n\qquad n=0,\pm 1,\pm 2,\ldots.\ \square$$

Theorem 16.6 Suppose that $\tilde{f}$ is a functional in Z' (or in S') with period X and Fourier coefficients $\{F_n\}$, and let $\tilde{g}$ be an indefinite integral of $\tilde{f}$: then it will follow that $\tilde{g}$ is of the form

$$\tilde{g}(x)=F_0x+\sum_{n=-\infty}^{\infty}G_n\exp(2\pi inx/X)$$

where the series converges in Z' (or in S'), and where

$$G_n=\frac{XF_n}{2\pi in}\qquad n=\pm 1,\pm 2,\ldots$$

with G_0 equal to an arbitrary constant: moreover if $F_0=0$ then $\tilde{g}$ will be periodic with Fourier coefficients G_n, and arbitrary G_0. □

Periodic functionals may be generated by sampling the transform of a functional, as follows.

Theorem 16.7 Suppose $\tilde{f}$ is a convolute in S' (or Z') and let $\tilde{F}$ be its Fourier transform: then $\tilde{F}$ will be equal to a multiplier $F(y)$ in S' (or D') and it will follow that, for each $X>0$,

$$X \sum_{n=-\infty}^{\infty} \tilde{f}(x-nX)$$

will converge in S' (or Z') to a periodic functional $\tilde{g}$ whose Fourier coefficients are

$$G_n = F(n/X) \qquad n=0, \pm 1, \pm 2, \ldots. \quad \square$$

The conditions in the above theorem may be relaxed, provided an unambiguous meaning can be given to the values of $F(y)$ at each point. It is not even necessary for $F(y)$ to be continuous, as the following shows.

Theorem 16.8 Suppose $\tilde{f} \in Z'$ has a transform $\tilde{F} \in D'$ that is regular and equal to a function F that is of bounded variation on each finite interval (though not necessarily on $(-\infty, \infty)$): then (i) $F(y)$ will be equal a.e. to a function $F_D(y)$ such that, at all y,

$$F_D(y) = \tfrac{1}{2}[F_D(y^-) + F_D(y^+)],$$

(ii) also

$$X \sum_{n=-\infty}^{\infty} \tilde{f}(x-nX) \tag{16.11}$$

will converge in Z' to define a periodic functional $\tilde{g}$ whose Fourier coefficients G_n are given by

$$G_n = F_D(n/X) \qquad n=0, \pm 1, \pm 2, \ldots,$$

and (iii) if in addition $\tilde{f} \in S'$ and $F(y)/(1+|y|)^N$ is of bounded variation on $(-\infty, \infty)$ for some $N>0$, then (16.11) will converge in S'. □

The above two theorems provide an abundant source of examples of periodic functionals. For instance, choosing $\tilde{f} \in \tilde{\delta}$ and $F(y)=1$ we have (in S'):

$$X \sum_{n=-\infty}^{\infty} \delta(x-nX) \leftrightarrow \sum_{n=-\infty}^{\infty} \delta(y-n/X). \tag{16.12}$$

Choosing $\tilde{f} = \delta'$ and $F(y) = 2\pi i y$ we obtain, again in S',

$$X \sum_{n=-\infty}^{\infty} \delta'(x-nX) \leftrightarrow X^{-1} \sum_{n=-\infty}^{\infty} 2\pi i n \delta(y-n/X). \tag{16.13}$$

Or again with $\tilde{f}=x^{-1}$ and $F(y)=-\pi i \operatorname{sgn} y$, (13.6), we obtain (in S')

$$X\sum_{n=-\infty}^{\infty}(x-nX)^{-1}\leftrightarrow-\pi i\sum_{n=1}^{\infty}[\delta(y-n/X)-\delta(y+n/X)]. \quad (16.14)$$

16.5 Products and convolutions

We start with convolutions of periodic functionals, for which the theorems are easier to state than is the case with products. This was the case also, in chapter 15, with ordinary Fourier series.

Theorem 16.19 Suppose that $\tilde{f}$ is a functional in Z' (or S') with period X, and that $\tilde{g}$ is a convolute in Z' (or S') and let the transforms be $\tilde{F}$ and $\tilde{G}$: then $\tilde{G}=G$ is a multiplier in D' (or S'), and

$$\tilde{h}=\tilde{f}*\tilde{g}$$

is a periodic functional in Z' (or S') with Fourier coefficients

$$H_n=F_nG(n/X). \quad \square$$

In order to convolute two periodic functionals we need to define a generalization of the T-convolution. This is achieved using the unitary functions.

Theorem 16.20 Let $\tilde{f}$ and $\tilde{g}$ be functionals of period X, both in Z' (or both in S'), and let φ be a unitary function in Z (or S) of parameter X: then

$$\tilde{h}=X^{-1}[\tilde{f}*(\varphi\tilde{g})] \quad (16.15)$$

defines a functional $\tilde{h}$ in Z' (or S') of period X whose Fourier coefficients are

$$H_n=F_nG_n \qquad n=0,\pm 1,\pm 2,\ldots,$$

where $\{F_n\}$ and $\{G_n\}$ are the coefficients of $\tilde{f}$ and $\tilde{g}$. $\square$

In this the functional $\tilde{h}$ may be called the T-convolution of $\tilde{f}$ and $\tilde{g}$, though it should be noted that (16.15) represents an ordinary convolution between $\tilde{f}$ and $\varphi\tilde{g}$ as defined in chapters 13 and 14.

We now consider products.

Theorem 16.21 Suppose $\tilde{f}$ is a functional in Z' (or S') of period X and that $\tilde{g}$ is a multiplier in Z' (or S'), and let $\{F_n\}$ be the Fourier coefficients of $\tilde{f}$ whilst $\tilde{G}$ is the Fourier transform of $\tilde{g}$: then it follows that

$$\tilde{f}\tilde{g}\leftrightarrow\sum_{n=-\infty}^{\infty}F_n\tilde{G}(y-n/X)$$

where the summation converges in D' (or in S'). $\square$

Finally we come to the products of two periodic functionals.

Theorem 16.22 Suppose $\tilde{f}$ is a functional in S' of period X, and that $g(x)$ is an ordinary function of period X that is everywhere infinitely differentiable: then g is a multiplier in S' and the product

$$\tilde{h}=\tilde{f}g$$

exists, where $\tilde{h}\in S'$ has period X, and the Fourier coefficients of $\tilde{h}$ are given by

$$H_n=\sum_{m=-\infty}^{+\infty} F_m G_{n-m} \qquad n=0, \pm 1, \pm 2, \ldots,$$

where the sum converges in the ordinary sense, and $\{F_n\}$ and $\{G_n\}$ are the Fourier coefficients of $\tilde{f}$ and $\tilde{g}$. □

If we wish to broaden the above to cover functionals in Z' we must correspondingly place conditions on g.

Theorem 16.23 Suppose $\tilde{f}\in Z'$ is of period X and has Fourier coefficients $\{F_n\}$, and that $g(x)$ is an ordinary function of period X that is everywhere infinitely differentiable and whose Fourier coefficients G_n are equal to zero for $|n|$ greater than some integer $N>0$: then g is a multiplier in Z' and the product $\tilde{h}=\tilde{f}g$ defines a functional $\tilde{h}\in Z'$ of period X whose Fourier coefficients are given by

$$H_n=\sum_{m=-N}^{N} G_m F_{n-m} \qquad n=0, \pm 1, \pm 2, \ldots. \quad □$$

BIBLIOGRAPHY

The literature on Fourier transformation is large, and the following is a very selective list of books and articles that the author has found particularly useful. Not every theorem in the text is to be found in an exactly equivalent form in the literature, but in all cases the background from which the theorem springs is described in the references cited.

Chapters 1–4. The basic level of mathematics presupposed is that of Jeffrey (1985) or Stephenson (1983). The applications and manipulation of Fourier theory are described in a non-rigorous way in Champeney (1985) and Stuart (1961) at an elementary level, or in Bracewell (1978), Champeney (1973) and Papoulis (1962, 1968, 1981) at a more advanced level. A fairly extensive tabulation of Fourier transforms is in Erdélyi (1954). Further background in mathematical analysis (excluding Lebesgue integration) appears in Apostol (1974), and the handbook (Howson, 1972) is very useful. Lebesgue integration, based on step functions, is clearly described in Weir (1973), whilst the approach based on measure appears in Titchmarsh (1968): concise descriptions are also included in Dym and McKean (1972), Jones (1966) and Weiner (1951).

Chapters 5–10. Results on convolution kernels are to be found in books on Fourier series or transforms, such as Bochner (1959), Edwards (1979, 1982), Katznelson (1976), Titchmarsh (1962, 1968), Weiner (1951) and Zygmund (1959). The properties of good functions and their transforms usually receive special attention in the context of generalized functions, for example Jones (1966) and Lighthill (1959), and for the results in the complex plane see Gel'fand *et al.* (1964–9), volume I, and Zemanian (1965). The classic theory of transformation of functions in class L^p, chapter 8, appears in Titchmarsh (1962), and also many aspects are covered in Bochner (1959), Katznelson (1976) and in the final chapters of Zygmund (1959). However, not all these books incorporate the results of Carleson

(1966) and Hunt (1968) or of Beckner (1975). The material for chapter 9 comes mainly from the area of generalized functions (Gel'fand, 1964–9; Jones, 1966; Lighthill, 1959; Zemanian, 1965), though many results emerge from the material covered in Bochner (1959). For chapter 10 the positive-definite functions are described in Bochner (1959), and further references on the Gibbs phenomenon are given in Dym and McKean (1972), whilst the results on complex transforms originate from Titchmarsh (1962).

Chapter 11 is a summary of Wiener (1930) together with some material from Wiener (1951, 1964).

Chapters 12–14. The approach to generalized functions based on sequences is introduced in Lighthill (1959) and developed further in Jones (1966). The approach based on linear continuous functionals is described in Zemanian (1965), as well as in the original French in Schwartz (1966) and in the six volumes of Gel'fand (1964–9); volume I contains a good introductory survey. Generalized functions are also described in Shilov (1968), though with little coverage of Fourier transformation.

Chapters 15, 16. The literature on Fourier series is even larger than that on Fourier transforms. Elementary accounts based on Riemann integration appear in Jeffrey (1985) and Sneddon (1961), and at a more advanced level in Apostol (1974). Accounts based on the Lebesgue integral range from the small Hardy and Rogosinski (1962) up to the monumental classic Zygmund (1959). The use of generalized functions is included in Gel'fand (1964–9), Jones (1966), Lighthill (1959), Schwartz (1966) and Zemanian (1965). A modern perspective on the complete subject of Fourier series appears in Edwards (1979, 1982), which also contains a large bibliography and pointers toward modern research.

Abramowitz, M. and Stegun, I. A. (eds.) (1966) *Handbook of Mathematical Functions with Formulas, Graphs and Mathematical Tables*. New York: Dover

Apostol, T. M. (1974) *Mathematical Analysis*, 2nd edn. Reading, MA: Addison-Wesley

Beckner, W. (1975) *Ann. Math.* **102**, 159–82

Bochner, S. (1959) (translated by M. Tenenbaum and H. Pollard) *Lectures on Fourier Integrals*. Princeton University Press

Bracewell, R. N. (1978) *The Fourier Transform and its Applications*. Tokyo: McGraw-Hill

Carleson, L. (1966) *Acta Mathematica* **116**, 135–57

Champeney, D. C. (1973) *Fourier Transforms and their Physical Applications*. London: Academic Press

Champeney, D. C. (1985) *Fourier Transforms in Physics*. Bristol: Adam Hilger

Churchill, R. V. (1960) *Complex Variables and Applications*. New York: McGraw-Hill

Dym, H. and McKean, H. P. (1972) *Fourier Series and Integrals*. New York: Academic Press

Edwards, R. E. (1979) *Fourier Series: A Modern Introduction*, Vol. I. New York: Springer-Verlag

Edwards, R. E. (1982) *Fourier Series: A Modern Introduction*, Vol. II. New York: Springer-Verlag

Erdélyi, A. (ed.) (1954) Tables of Integral Transforms (Bateman Manuscript Project). New York: McGraw-Hill

Fourier, J. B. J. (1955) (translated by A. Freeman) *The Analytical Theory of Heat*. New York: Dover

Gel'fand, I. M. *et al.* (1964–9) (translated by E. Saletan) *Generalized Functions*, Vols 1–6. New York: Academic Press

Gradshtein, I. S. and Ryzhik, I. M. (1965) *Table of Integrals, Series and Products*. New York: Academic Press

Grattan-Guinness, I. (1972) *Joseph Fourier, 1768–1830*. Cambridge, MA: MIT Press

Hardy, G. H. and Rogosinski, W. W. (1962) *Fourier Series*. Cambridge University Press

Herivel, J. (1975) *Joseph Fourier:* The Man and the Physicist. Oxford: Clarendon Press

Howson, A. G. (1972) *A Handbook of Terms used in Algebra and Analysis*. Cambridge University Press

Hunt, R. A. (1968) in *Orthogonal Expansions and their Continuous Analogues* (Proc. Conf., Edwardsville, IL), pp. 235–55. Carbondale: Southern Illinois University Press

Jeffrey, A. (1985) *Mathematics for Scientists and Engineers*. Van Nostrand Reinhold

Jones, D. S. (1966) *Generalized Functions*. London: McGraw-Hill

Katznelson, Y. (1976) *An Introduction to Harmonic Analysis*. New York: Dover

Khintchine, A. (1934) *Math. Annalen*. **109**, 604–15

Kramers, H. A. (1927) *Estratto dagli Ati del Congresso Internazionale di Fisici Como* (Nicolo Zonichelli, Bologna)

Kronig, R. (1942) *Ned. Tijdschr. Natuurk* **9**, 402

Lighthill, M. J. (1959) *Introduction to Fourier Analysis and Generalized Functions*. Cambridge University Press

Papoulis, A. (1962) *The Fourier Integral and its Applications*. New York: McGraw-Hill

Papoulis, A. (1968) *Systems and Transforms with Applications in Optics*. New York: McGraw-Hill

Papoulis, A. (1981) *Circuits and Systems: A Modern Approach*. New York: Rinehart & Winston

Pitts, C. G. C. (1972) *Introduction to Metric Spaces*. Edinburgh: Oliver and Boyd

Schwartz, L. (1966) *Théorie des Distributions*. Paris: Hermann

Shilov, G. E. (1968) (translated by B. D. Seckler) *Generalized Functions and Partial Differential Equations*. New York: Gordon & Breach

Sneddon, I. N. (1961) *Fourier Series*. London: Routledge and Kegan Paul

Stephenson, G. (1983) *Mathematical Methods for Science Students*. Longman

Stuart, R. D. (1961) *An Introduction to Fourier Analysis*. London: Science Paperbacks

Sutherland, W. A. (1975) *Introduction to Metric and Topological Spaces*. Oxford: Clarendon Press

Titchmarsh, E. C. (1962) *Introduction to the Theory of Fourier Integrals*. Oxford: Clarendon Press

Titchmarsh, E. C. (1968) *The Theory of Functions*. Oxford University Press

Weir, A. J. (1973) *Lebesgue Integration and Measure*. Cambridge University Press
Wiener, N. (1930) *Acta Math.* **5**, 117–258
Wiener, N. (1951) *The Fourier Integral and Certain of its Applications*. New York: Dover
Wiener, N. (1964) *Selected Papers*. Cambridge, MA: MIT Press
Wooten, F. (1972) *Optical Properties of Solids*. New York: Academic Press
Zemanian, A. H. (1965) *Distribution Theory and Transform Analysis*. New York: McGraw-Hill
Zygmund, A. (1959) *Trigonometric Series*, Vol. 1–2. Cambridge University Press

INDEX

Numbers refer to page numbers. Numbers in italics refer to pages containing definitions or other basic reference material. Abbreviations and symbols for mathematical entities are included under the appropriate letter of the alphabet when the choice of letter is unambiguous, whilst other symbols applicable generally to any function appear under the letter 'f'.

Printed in the United States
By Bookmasters